in_focus

Seeds
that give

PARTICIPATORY PLANT BREEDING

in_focus

IDRC's *In_Focus* Collection tackles current and pressing issues in sustainable international development. Each publication distills IDRC's research experience with an eye to drawing out important lessons, observations, and recommendations for decision-makers and policy analysts. Each also serves as a focal point for an IDRC Web site that probes more deeply into the issue, and is constructed to serve the differing information needs of IDRC's various readers. A full list of *In_Focus* Web sites may be found at **www.idrc.ca/in_focus**. Each *In_Focus* book may be browsed and ordered online at **www.idrc.ca/booktique**.

IDRC welcomes any feedback on this publication.

Please direct your comments to The Publisher at **pub@idrc.ca**.

in_focus

Seeds
that give

PARTICIPATORY PLANT BREEDING

by **Ronnie Vernooy**

INTERNATIONAL DEVELOPMENT RESEARCH CENTRE
Ottawa • Cairo • Dakar • Montevideo • Nairobi • New Delhi • Singapore

Published by the International Development Research Centre
PO Box 8500, Ottawa, ON, Canada K1G 3H9
http://www.idrc.ca

© International Development Research Centre 2003

National Library of Canada cataloguing in publication data

Vernooy, Ronnie

Seeds that give : participatory plant breeding

ISBN 1-55250-014-4

1. Plant breeding.
2. Plant breeding — Case studies.
3. Field crops — Breeding.
4. Agrobiodiversity conservation.
I. International Development Research Centre (Canada)
II. Title.

SB123.V47 2003   631.5'2   C2003-980109-8

IDRC Books endeavours to produce environmentally friendly publications.
All paper used is recycled as well as recyclable. All inks and coatings are
vegetable-based products.

This book is also an integral part of IDRC's thematic Web dossier on
participatory plant breeding: **http://www.idrc.ca/seeds**. The full text of
the book is available online and leads the reader into a virtual web of
resources that explores a decade of research on agrobiodiversity and PPB.

Contents

Preface

Diversity means life; diversity means choice. Unfortunately, around the world the spaces for the maintenance and creation of (new) diversity are becoming more and more confined. Biological diversity, in environments increasingly disturbed by human intervention, is under serious threat. Globalization forces are imposing limits on the ways people shape and reshape socio-economic, cultural, and political diversity.

At the same time, in many places efforts are underway to maintain or open up new room for the appreciation, use, and further evolution of diversity. In 1992, following the United Nations Conference on Environment and Development (UNCED or the

"Earth Summit"), staff at Canada's International Development Research Centre (IDRC) developed a program to support these efforts. IDRC's biodiversity program was born to put and keep biodiversity high on the agenda of research and development organizations in the South, in Canada, and around the globe. In 1997, the biodiversity program evolved into the Sustainable Use of Biodiversity (SUB) program initiative, retaining its major objectives and approach:

- **to promote** the use, maintenance, and enhancement of the knowledge, innovations, and practices of indigenous and local communities to conserve and sustainably use biodiversity;

- **to develop** incentives, methods, and policies that facilitate the development of strategies for the conservation and enhancement of *in situ* agricultural and aquatic biodiversity; and the participation of communities in their design and implementation; and

- **to support** the creation of policies and legislation that recognize the rights of indigenous and local communities to genetic resources and to the equitable sharing of benefits of the use of these resources.

This In_Focus book presents fragments of the arduous biodiversity research work carried out and ongoing in numerous, often far away and little known places around the world. The book builds on an internal review of 10 years of IDRC support to agricultural biodiversity. Both the book and the review aim to take stock of the cumulative efforts in terms of research and development achievements and challenges. They are meant to be formative: to enhance ongoing work, conceptually, methodologically, and practically.

Over the last 10 years I have had the great privilege to interact with and learn from the researchers, farmers, extension agents, and government officials who carry out or support the participatory plant breeding efforts described here. I thank them all for trying out unwalked paths and showing the way.

Salvatore Ceccarelli, Noemi Espinoza, Sanjaya Gyawali, Humberto Lambrada Ríos, Yiching Song, and Louise Sperling have given extra voice to my story. I acknowledge their enlightening contributions. Marcel Vernooij and Louise Sperling, in their roles as agricultural and biodiversity policy director and participatory plant breeding research manager, respectively, were so kind to respond in very clear words to the key question: "What decisions do you make in your day to day work concerning agrobiodiversity?" I hope that the recommendations presented here to guide decision-making will be positively received by them and by their colleagues.

Numerous IDRC colleagues over the years have shaped and reshaped the biodiversity and SUB programs. Without their strong commitment, innovative ideas, and constructive critiques, IDRC's support for participatory plant breeding would never have made it to its 10th anniversary!

Bob Stanley, accomplished wordsmith, accepted to embark on the writing train with me. His craftsmanship can be found throughout this book. I am grateful for his invaluable contribution, and for his endurance to process the never-ending stream of corrections.

Bill Carman and the IDRC Communications team had the courage to review the manuscript and provided excellent feedback. They also took care of the production and publication process. It has been a pleasure to work with them.

The pioneering efforts highlighted in this book to maintain or enlarge the space for the dynamic evolution of diversity, for the improvement of agricultural productivity, and for the recognition of both farmers' and breeders' knowledge and skills are a great inspiration. I hope that these efforts and the new initiatives building on them will receive the strong and much needed support from key decision-makers in research and policy.

This book is dedicated to the memory of Marie Béatrice Dubé.

Ronnie Vernooy
January 2003

Ronnie Vernooy is a senior program specialist at the International Development Research Centre, Ottawa, Canada. Trained as a rural development sociologist, his interests include farmer experimentation and organization, natural resource management, agricultural biodiversity, and participatory (action) research methods including monitoring and evaluation. His current work focuses on Southeast Asia, Central America, and Cuba. He has a special interest in Nicaragua, where he carried out field research in both hillside and coastal environments during 1985–86, 1988–91, and 1997–98. Recent publications include *Taking care of what we have: participatory natural resource management on the Caribbean coast of Nicaragua* (editor and coauthor, IDRC/CIDCA-UCA 2000), *Para una mina de oro se necesita una mina de plata: historiando sobre la Costa Caribe de Nicaragua 1910-1979* (CIDCA-UCA, 2000), and *Voices for change: participatory monitoring and evaluation in China* coeditor, IDRC/YSTP 2003).

Part 1

The Issue

Mega-crops and genetic erosion

The sign beside the highway in rural Canada reads "If you ate today thank a farmer." Perhaps it should also say "thank a plant breeder," because most people in the North — as well as a large percentage of those in the South — eat today thanks to the remarkable advances in agricultural science and technology.

It is agricultural science that has enabled us to defy the pessimistic projections of Malthus and continue to feed an ever-increasing world population. Although the rate of increase has slowed in the past generation, the tide of humanity continues to rise: today 6 billion, soon 8 billion, perhaps as many as 10 billion by the year 2050. But here is a disconcerting fact: as the human population continues to expand, the number of crops on which most of us depend for sustenance is shrinking.

No one knows exactly how agriculture began 8 000 to 10 000 years ago; how our ancestors first began to identify, manipulate, and manage certain wild plants and creatures as a source of food. We do know that the invention of agriculture represented a sea change in the evolution of humanity leading to the social systems and structures that we call "civilization."

Over the millennia, the processes of farmer experimentation led to the domestication of an ever wider range of plants to meet specific needs, preferences, and environmental conditions. The result was thousands of different and genetically unique plant varieties cultivated in farming systems. Yet today only about 150 plant species are cultivated. Twelve of these provide three-quarters of the world's plant-based food, while fully half the world's plant-based food supply comes from a limited number of varieties of just a few plant species. These are the "mega-crops": rice, wheat, maize, as well as sorghum and millet, potatoes, and sweet potatoes.

Mega-crops — the high-yielding, high-input plants developed by scientists at the international agricultural research centres (IARCs) around the world. These were the foundation of what became known as the Green Revolution, which expanded agricultural output exponentially in many developing regions and provided food for hundreds of millions of people.

Yet at the heart of this success lies a threat. The top-down system of agricultural research, where farmers are seen merely as recipients of research rather than as participants in it, has contributed to an increased dependence on a relatively few plant varieties. This trend and the increasing industrialization of agriculture are key factors in what can only be called "genetic erosion." The term refers to both the loss of species and the reduction of variety, and it includes not only plants but also animals and microorganisms, as well as the gradual breakdown of the processes that maintain the evolution of diversity. These processes include the constantly evolving knowledge, innovations, practices, and forms of organization of farmers in local and indigenous communities. The practices relating to the production, harvesting, and preparation of food are often integral to peoples' cultural identities.

Farmers' knowledge about agricultural diversity continues to be crucial in many places, but their crops and cropping systems are also under increasing pressure. The Food and Agricultural Organization of the United Nations (FAO) estimates that of the roughly quarter-of-a-million plant varieties available to agriculture, only about 7 000 — less than 3% — are in use today. With disuse comes neglect and, possibly, extinction.

> According to the FAO, replacement of local varieties with improved or exotic varieties, or both, is the major cause of genetic erosion around the world.

Why diversity matters

Modern agriculture is like a huge inverted pyramid; it rests on a precariously narrow base. Genetic erosion could threaten the future food supply if anything should happen to reduce the effectiveness of the high-yielding varieties that we have come to depend on.

There's the first paradox: the very success of agricultural science has led to a concentration on a small number of varieties designed for intensive agriculture, and a dramatic reduction in the diversity of plant varieties available for continued agricultural research and development.

In the past, researchers have been able to depend on farmers to retain sufficient crop diversity to provide the "new" genetic material they need. Crop breeders tend to rely increasingly on a narrow set of improved varieties. Homogeneous modern agriculture threatens the source of genetic diversity, and thus threatens both local and global food security.

It must be said also that the success of the plant breeders has not been totally unblemished. High-yielding varieties are frequently also high-maintenance varieties. They usually require regular applications of fertilizer and other inputs. In other words, they do not thrive on poor soils or in adverse conditions.

These constraints effectively put such high-yielding varieties beyond the reach of millions of small-scale farmers, who cannot afford the high-priced seed and fertilizer. Worse, most of these farmers reject the plant breeders' offerings because they simply are not designed for marginal farmland — they meet neither the farmer's needs nor local preferences.

Yet these resource-poor farmers, a large number of whom are women, produce as much as 20% of the world's food. About one-quarter of the world's population depends on these marginal lands for their food.

Farmers under these conditions typically employ mixed farming techniques, growing both grains and vegetables, raising a few chickens for eggs and meat, and, if they can afford it, keeping a few animals — pigs, goats, a cow or two. They select and plant seed from their own crops, and they exchange seed with neighbours or family members. Seeds are sometimes given as valued gifts. For many it is a subsistence living, often subsidized by work off the farm. But in a good season there may be a surplus to take to market.

Now the second paradox: the key to increasing biological and cultural diversity may lie with these traditional, small-scale farmers. For in their struggle simply to survive on poor soils with limited resources, small farmers continue to allow plant varieties to evolve. They select plant types (rather than varieties) based on their own observations and according to their specific needs. For example, local conditions may favour a shorter, sturdier plant; or the flavour, even the colour of the end product may be important.

The result is that to a surprising extent these farmers have become custodians of diversity. Through their skills as plant breeders — based on experience and observation rather than on scientific knowledge — they are maintaining the genetic variation that is essential to the continued evolution and adaptation of plant genotypes. They also bring to the process a broad cultural diversity expressed through local knowledge, language, practices, and forms of organization that are equally important in conserving biodiversity.

Rethinking conventional breeding strategies means above all recognizing the key roles of farmers.

Dynamic conservation and improvement

The one-size-fits-all approach to plant breeding not only fails to meet the needs of small farmers in the developing world, it also contributes to the loss of agricultural diversity, or agrobiodiversity. The loss of agrobiodiversity in turn leads to a reduction in the capacity of agricultural ecosystems to continue producing renewable resources. It also limits the ecosystem's ability to deal with change, which leads to decreased resilience. In short, it's a downward spiral. In the words of the FAO's 1998 report on the state of the world's plant genetic resources for food and agriculture: "It may be necessary to rethink conventional breeding strategies."

Rethinking conventional breeding strategies means above all recognizing the key roles of farmers and their knowledge and social organization in the management and maintenance of agrobiodiversity. Recognizing these roles is the basis of an approach to agricultural research known as participatory plant breeding, or PPB. **Simply stated, the aim of PPB is to ensure that the research undertaken is relevant to the farmers' needs.** Researchers work directly with the farmers, and much of the testing takes place on the farm.

In PPB, instead of playing a supporting role in the research, the farmers are treated as partners in the undertaking. In fact, the farmers will often take the lead, sometimes combining their own seeds with the material supplied by the plant breeders. Because the farmers' varieties are well adapted to local conditions, the results are more likely to meet with approval. And when that happens, the farmers don't hesitate to start multiplying and distributing the seed. **It is a dynamic process of conservation and improvement.**

PPB and the *in situ* conservation of agrobiodiversity — which means maintaining the diversity of plant species on farms in the habitats where they originated and continue to evolve — are two complementary methodologies. Small farmers breed their own improved varieties simply to survive. In doing so they maintain that diversity, but they do not distinguish between conservation and development. PPB is an approach that promotes development while conserving diversity.

PPB empowers small farmers and validates the logic behind their choices. It gives the farmers a greater measure of control over their livelihood, and for those living at or near subsistence level it provides the opportunity to break out of the cycle of poverty. Perhaps no group benefits more from the PPB approach than poor rural women. It is the women who provide much of the farm labour, process and store grains and other crops, and prepare the food. Because in many places they also preserve the best seed for planting, they play a key role in managing plant genetic resources.

Here is the third paradox: the countries that are richest in genetic material are often the poorest in terms of economic wealth. Many of the crops on which the world depends today originated in what we now call the developing world — potatoes from the Andes in Latin America, wheat from West and Central Asia, for example. Not surprisingly, the greatest genetic diversity is still to be found in these regions, as illustrated in Figure 1.

If that diversity is to be preserved for the future food security of all humanity, then ways must be found for the people of these regions, who are in effect its custodians, to finally share in the benefits. Thus PPB must also deal with the sensitive issue of farmers' rights. This is a concept that has already been adopted by many advocates of PPB and is implicit in the Convention on Biological Diversity (CBD), which calls for a fair and equitable sharing of benefits arising out of the use of genetic resources. The concept goes beyond just compensating farmers for their role in conserving and improving plant genetic resources. It extends to enabling communities to gain more control over their own biomaterials, sharing of knowledge and technology, capacity building, and access to land and markets.

> Many researchers see participatory plant breeding as an essential step to securing the world's food supply.

A decade of research

There are many approaches to PPB. Some development organizations see it as a means of alleviating poverty and increasing the food supply in some of the world's poorest regions. Others promote it as a way of making research less costly and more efficient. Still others focus on issues such as farmers' rights and greater equality for women. Many researchers see it as an essential step to securing the world's food supply. Since 1992,

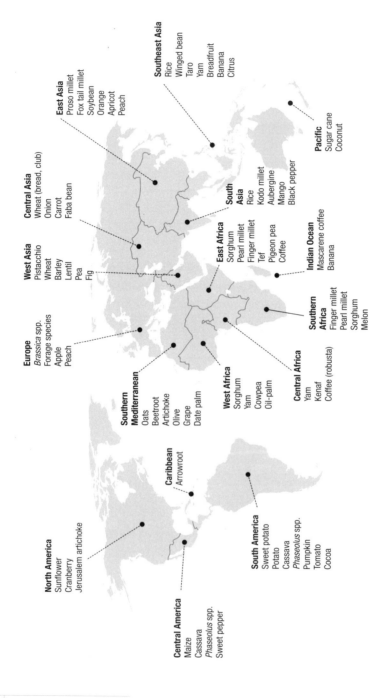

North America
Sunflower
Cranberry
Jerusalem artichoke

Central America
Maize
Cassava
Phaseolus spp.
Sweet pepper

Caribbean
Arrowroot

South America
Sweet potato
Potato
Cassava
Phaseolus spp.
Pumpkin
Tomato
Cocoa

Europe
Brassica spp.
Forage species
Apple
Peach

Southern Mediterranean
Oats
Beetroot
Artichoke
Olive
Grape
Date palm

West Asia
Pistacchio
Wheat
Barley
Lentil
Pea
Fig

Central Asia
Wheat (bread, club)
Onion
Carrot
Faba bean

East Asia
Proso millet
Fox tail millet
Soybean
Orange
Apricot
Peach

West Africa
Sorghum
Yam
Cowpea
Oil-palm

Central Africa
Yam
Kenaf
Coffee (robusta)

East Africa
Sorghum
Pearl millet
Finger millet
Tef
Pigeon pea
Coffee

Southern Africa
Finger millet
Pearl millet
Sorghum
Melon

South Asia
Rice
Kodo millet
Aubergine
Mango
Black pepper

Indian Ocean
Mascarene coffee
Banana

Southeast Asia
Rice
Winged bean
Taro
Yam
Breadfruit
Banana
Citrus

Pacific
Sugar cane
Coconut

Figure 1. **Regions of diversity of major cultivated plants (adapted from FAO 1998).**

Canada's International Development Research Centre (IDRC) has maintained a particular focus on research aimed at supporting the conservation of biodiversity. Today that focus continues through IDRC's program for the Sustainable Use of Biodiversity (SUB). Underlying this effort is a body of applied research in the fields of agriculture, fisheries, forestry, nutrition, and health supported by IDRC during the 1970s and 1980s.

This book provides a brief examination of a decade of support for research targeted directly or indirectly at the field of participatory plant breeding. The accumulated results of this research represent a body of knowledge and experience worth sharing. It begins with a review of the approach and key research questions, illustrated by brief reports on six diverse projects around the world. Next it examines accumulated project results in the light of expected outcomes. There follows a series of recommendations for future activities, based on the lessons learned over the past decade. The book concludes with a speculative look at future directions for research and policy on PPB as an integral part of a global agrobiodiversity agenda.

Part 2

The Approach

A focus on community

Agrobiodiversity is complex and many-faceted, but ultimately it's all about food. More food, better food, secure food, food for all. However, if we are to avoid the mistakes of the past, the focus must be on people, on the communities where diversity lives, and on engaging the people of these communities in a broader approach to the sustainable use of agrobiodiversity.

There is a scientific rationale for such a focus: it is that these communities are the principal stewards of the greater share of the world's biodiversity. Communities contain the skills and knowledge that have contributed to the wide range of distinct types and varieties of plants, animals, fish, and microorganisms that are vital to their (and our) food and health security.

Communities have shaped and maintained the environments that support this diversity.

From the Andean highlands to the foothills of the Himalayas to the arid zones of Africa and the Middle East, farmers who have never heard the word "agrobiodiversity" understand full well the importance of selecting and preserving the best seed, the sturdiest stock. Farmers and the people of rural communities — the women and men who are the stewards — value agrobiodiversity for their very survival as well as for a variety of sociocultural reasons. Other users with a stake in agrobiodiversity are concerned with ecological, economic, and political values. So it is essential that any concerted effort to strengthen global agrobiodiversity begin by identifying what agrobiodiversity means — and to whom.

Agrobiodiversity is a broad concept that includes a variety of biological diversity components — from agricultural ecosystems, to crop varieties, to genes in plant and animal species. From an ecological perspective, agrobiodiversity supports and protects human lives, it provides continued inputs for evolution; it increases the productive capacity of ecosystems. Lessen agrobiodiversity and you weaken the resilience of the system and its capacity to deal with change. When this happens, communities face more limited options in managing their land and resources. And the end result is that opportunities for the creation and re-creation of farmer knowledge and experimentation — the very processes that are essential for agrobiodiversity conservation, evolution, and improvement — are lost.

Stemming the loss of diversity is critical. Community participation is the key. The most effective way to address the loss of agrobiodiversity is to concentrate first on the importance of biodiversity resources to the livelihoods of people, particularly in marginalized communities. It is in these marginal environments where resource conflicts and competition are often greatest. Second, these communities must be given the opportunity to participate as equals in enhancing the sustainable use of agrobiodiversity.

From that twofold focus comes the IDRC approach to PPB research — an approach that is driven by three specific objectives:

→ **promote** the use, maintenance, and enhancement of the knowledge, innovations, and practices of indigenous and local communities to conserve and sustainably use biodiversity;

→ **develop** incentives, methods, and policies that facilitate the development of strategies for the conservation and enhancement of *in situ* agricultural biodiversity, and the participation of communities in their design and implementation; and

→ **support** the creation of policies and legislation that recognize the rights of indigenous and local communities to genetic of resources and to the equitable sharing of benefits of the use of these resources.

By no means is this an isolated view. The approach and the objectives have a global context that builds directly on Article 12 of the CBD, which deals with research and training. They also link to many other CBD articles, such as those dealing with conservation and sustainable use, *in situ* conservation, access to genetic resources, and technical and scientific cooperation — to name a few — as well as to other international agreements.

PPB requires close collaboration between researchers and farmers, and potentially other stakeholders, to bring about plant genetic improvements.

By this point it should be clear that the issue of agrobiodiversity conservation is far from simple. Methodologies that combine understanding of the biological and the social aspects of complex systems are urgently needed. They should also utilize linkages between local and broader economic and policy frameworks. PPB is one such methodology. It is an approach requiring close collaboration between researchers and farmers, and potentially other stakeholders, to bring about plant genetic improvements. This covers the complete cycle of research and development activities associated with plant genetic improvement, including

→ identifying breeding objectives,

→ generating genetic variability or diversity,

→ selecting within variable populations to develop experimental materials,

→ evaluating these materials (this is known as participatory variety selection, or PVS),

→ releasing materials, and

→ diffusion, seed production, and distribution.

The approach could also include assessing existing policy or legislative measures, or both, and designing new ones where needed. Farmers and breeders, and other stakeholders — such as traders, processors, distributors, and consumers — can take on different roles at various points in the cycle, but they join forces to bring about change.

The right questions

Research begins with questions, and getting the questions right is half the work. PPB research combines not just plant genetics and plant pathology, it also includes economics and elements of anthropology, sociology, marketing, production, and of course farmer know-how. Defining the questions, then, is no simple task, and it is only the beginning. The real challenge is to find answers that are appropriate in the context where the research is conducted; answers that perhaps can also be applied successfully in other contexts.

The agrobiodiversity and PPB research projects that are described in the next chapter of this book all attempt in one way or another to find answers to three types of questions:

Questions dealing with the knowledge and practice of in-situ conservation and improvement of agricultural genetic resources:

➤ What do farmers know about the properties and uses of agricultural genetic resources, including conservation and improvement? How can we ensure that this knowledge is respected and used **appropriately and fairly** for the benefit of both local communities and the wider society?

➤ These are times of agroecological and socioeconomic change. What are **viable** management practices, **fair** cost- and benefit-sharing mechanisms, and **useful** incentives to strengthen *in situ* conservation and improve agricultural genetic resources under changing conditions?

➤ How can we encourage new participatory models in agricultural genetic conservation and improvement — models that generate **mutual benefits**, encourage farmer-to-farmer exchange, and strengthen linkages between formal sector research-and-development and farmer experimentation?

Questions dealing with participation:

➤ What can be done to stimulate a more **meaningful participation** in research, development, and policy-making by the custodians of agricultural biodiversity?

➤ Are there **enabling** political and legislative conditions or changes that could help to make this happen?

Questions dealing with access, ownership, and compensation:

➤ What about policy or legislative changes? Are new regulations, incentives, or laws required to give marginalized farmers more **equitable access** to information and resources for conservation and improvement of agrobiodiversity?

→ What impacts do intellectual property rights (IPR) in living organisms have on farmers' experimentation and innovation? And do farmers have **fair access** to the benefits derived from these processes?

→ How can we work out fair access and compensation arrangements among stakeholders before actual interventions?

These questions represent a dynamic approach to crop conservation and improvement that incorporates the use of a PPB research methodology through the inclusion of elements such as social or gender analysis, consideration of policy linkages, and the formulation of policy and legislative alternatives.

Seeking the answers

As one of a number of development organizations supporting research on PPB and agrobiodiversity around the world, IDRC has accumulated a base of findings that is both broad and diverse. Many of the elements discussed above can be found in the design of the research that produced these results, and in many cases is still ongoing.

A decade of agrobiodiversity and PPB research has been more or less evenly spread across Africa, Asia, Latin America and the Caribbean, and the Middle East, and has included a number of global projects with research sites on several continents. Much of the research has been carried out in collaboration with other organizations. These include centres affiliated with the Consultative Group for International Agricultural Research (CGIAR), non-government organizations (NGOs), national agricultural research systems (NARS), and universities.

The research projects are as diverse as the participating organizations. They include all three major crop propagation types, with a clear focus on the major staple crops: rice, beans, maize, and to a lesser degree sorghum. Many projects focus on two or more crops, including combinations of open-pollinated, self-pollinated, and clonally reproduced, as well as vegetables, fruits, and other crops. Multicrop farming systems and home gardens are also included.

Most of the research has been conducted in unfavourable agro-ecosystems. "Unfavourable" in this context means agroecological areas with harsh climatic conditions, poor soils, and rugged land-scape, or any combination of these. Some work is also underway in sites that offer more favorable environments, for purposes of comparison of results and to explore how PPB techniques might be useful under better conditions. However, this book focuses on PPB research in unfavourable conditions.

Not surprisingly, productivity or diversity, or both, are the focus of at least half the projects. However, empowerment is also an element in a good number of projects. Empowerment? Not just a buzzword in this context, it has to do with making farmers — or even entire communities — real partners in research. It also means improving the technical expertise of farmer–breeders, and affirming the rights of local people to have control over, and ben-efit from, their genetic resources.

> Empowerment means making farmers real partners in research, and affirming the rights of local people to have control over, and benefit from, their genetic resources.

The participatory approach can be either consultative or collabo-rative. In the former, farmers and other stakeholders are con-sulted by the researchers from the formal system, but have little or no direct influence on the project and no decision-making power to direct the project in one way or another. In other words, although the project is participatory, the researchers still keep research decision-making firmly in their own hands. More recently it is encouraging to see that a number of projects have gradually moved to a more collaborative approach, with researchers and farmers discussing research steps and sharing the decision-making on an equal footing.

User-differentiated analysis, particularly where it relates to gender, is now generally accepted as a very important feature of sound agricultural research. This type of analysis examines outcomes on the basis of a range of factors that may account for social differentiation, unequal power relationships, or economic inequity. These include gender, age, education, class, caste, and ethnicity. For example, it has been found that women make seed selections based on quite different factors than those used by men. Also, different ethnic groups within a region can show distinct preferences for certain food characteristics, such as flavour or cooking quality.

Policy analysis is also an important component of many crop-improvement projects. For example, government policies relating to pricing and marketing may be a contributing factor in how farmers manage their systems. Other policy issues include intellectual property rights and the certification of organic produce — an increasingly important issue as the global demand for organic foods increases.

An understanding of the relationship between agrobiodiversity and the people who use, nurture, and depend upon it is vital to encouraging interactions that enhance and maintain this diversity. In Part 3 of this book we offer six participatory plant breeding "stories" — examples of projects that are helping to build that understanding. These are projects that are also innovating research practice with the aim of producing food and seeds for all.

Experiences from the Field

The six project "stories" that make up this chapter provide an overview of the many agrobiodiversity and PPB projects supported by IDRC over the past decade. They were selected to represent as wide a cross-section of the research as possible in terms of cropping systems, research objectives, types of participation, and methodological scope. As well, the work was carried out by centres affiliated with the CGIAR, by NARS, and by NGOs. In terms of geography, the projects cover Asia, Africa and the Middle East, and Latin America, as well as one global program. Time-wise, they include longer running efforts as well as work initiated more recently. All six initiatives are ongoing. Longer descriptions have been published by IDRC and are available in print and online at www.idrc.ca/seeds.

Participatory barley improvement in North Africa and the Middle East

Key Research Elements

Cropping system	Barley
Objectives (prioritized)	Productivity / diversity / empowerment
Participation	Researcher-led, collaborative
Social analysis	Gender
Policy component	Increasingly prominent (research policies, plant breeding, variety release)

A new way to work with farmers in dry areas

In many parts of North Africa and the Middle East, yields of key crops such as barley (a self-pollinated crop) are chronically low, and crop failures are common. Malnutrition is widespread in the poorest regions, and famine is a constant threat. Conventional breeding programs aimed at improving the crop have had little effect, largely because most farmers refuse to adopt the new varieties.

The conventional approach has been a centralized, top-down approach that pays little regard to the actual conditions that farmers face. What if you decentralized the breeding program, involved farmers right from the start, had breeders and farmers work side by side to learn from each other, and paid close attention to what the farmers told you? Revolutionary perhaps, but it's a revolution that has produced positive results.

In the late 1990s, a team of researchers at the International Centre for Agricultural Research in the Dry Areas (ICARDA) pioneered a new way to work with farmers in the marginal rainfall environments of Morocco, Syria, and Tunisia. Funded by IDRC, Italy, and BMZ/GTZ-Germany, the initiative brought together farmers and breeders with the common goal of fulfilling the needs of people living and working in the harsh conditions of the region.

In Syria, for example, "host farmers" in nine communities were linked with two research stations. These host farmers and their neighbours took care of the trials, which involved experimental lines from the research station and the farmers' own varieties. Farmers and breeders assessed the results independently in successive trials from 1997 to 1999. Several promising new varieties were identified from these trials.

Decentralized selection in farmers' fields avoids the risk of useful lines being discarded because of their relatively poor performance at experimental stations, where conditions are almost certainly more favourable.

It quickly became apparent that the farmers' selection criteria, largely based on environmental factors, were quite different from those used by the national breeding programs. To the surprise of many, the selections made by the farmers were at least as effective as those made by the breeders. Yields increased in areas where plant breeding had not previously been successful. Seeing these results, breeders quickly adopted new ideas and attitudes, becoming supporters of the participatory approach.

Conclusion: earlier plant breeding programs were ineffective on marginal lands because they seldom included among their selection criteria those traits that are important to farmers.

Corollary: decentralized selection in farmers' fields avoids the risk of useful lines being discarded because of their relatively poor performance at experimental stations (where conditions are almost certainly more favourable, through fertilization or irrigation for example). **Decentralized selection combined with farmer participation from the initial stages of the breeding process is a powerful methodology to fit crops to specific biophysical and socioeconomic contexts, and to respond to farmers' needs and knowledge.**

Addressing farmers' needs and interests

The researchers learned a number of other critical lessons from the project— among them, the fact that farmers can handle a large number of lines or populations, or both. Most notably, in Syria in phase 2 of the work, the number of lines assessed increased from around 200 up to 400 (Table 1). In fact, farmers warmly welcomed the ability to select among a large number of lines. Some farmers have started to produce seeds of the selected material. These seeds are being shared with other farmers, thereby reducing dependence upon seeds delivered by the breeders. This is leading to a more dynamic breeding process, with new materials regularly being passed from farmer to farmer.

Table 1. Quantity of breeding material in ICARDA's PPB research.

Country	Villages	Lines	Plot size (m²)	Farmers per village
Egypt	8	60	6	5
Eritrea	3	155	3	10–12
Jordan	6	200	6	5–12
Morocco	6	30–210	4.5	6–15
Syria (phase 1)	9	208	12	5–9
Syria (phase 2)	8	200–400	12	6–11
Tunisia	6	25–210	4.5	10–20
Yemen	3–6	100	3	15–20

Source: Ceccarelli (2000).

The researchers also noted that women's selection criteria often differed from men's, highlighting the importance of ascertaining when and why they differ. And they noted that farmers became empowered by their involvement in PPB, gaining the confidence to take decisions on crosses as well as on factors such as plot size and the number of locations.

Perhaps of equal importance to the researchers themselves, the project revealed the need for specific training in areas such as experimental design and data analysis suitable for situations where the environment (a farmer's field under farmer's management) cannot be under the scientists' control as it is in the research stations.

Building on success

So successful has been this pioneering approach that farmers have requested breeders to work with them using a similar approach to improve other crops. It has also spread to other countries in the region. ICARDA currently supports PPB programs on barley in Egypt, Eritrea, Jordan, and Yemen. In Bangladesh, Syria, Turkey, and Yemen, the same approach is being applied to research on lentils. Complementary to the PPB efforts, ICARDA has begun participatory research in natural resource management, in particular on sustainable land management in dry areas.

In each country, the success has been repeated. In Yemen, for example, a project that began with just three villages in the northern highlands quickly doubled to include three more villages in the central highlands. And the participatory approach has been used as a model in other projects carried out by the Agricultural Research and Extension Authority (AREA), ICARDA's national partner research organization. Another example: agricultural research authorities in Jordan have started to transform the national barley-breeding program into a decentralized, participatory program, and to extend PPB on bread and durum wheat.

Local agricultural research committees in Latin America

Key Research Elements

Cropping system	Beans, maize, cassava, potato, fruits
Objectives (prioritized)	Productivity / empowerment / diversity
Participation	Farmer-led, collaborative
Social analysis	Variable
Policy component	Not prominent

Experimenting and learning together

In the North, sending a problem to a committee is too often a means of avoiding action. Not so among the farmers and researchers of Latin America — here the committee has evolved as a platform for evaluating, adapting, and disseminating new technology. In addition, the committee has become an engine for rural development initiatives such as the formation of credit and marketing groups. Local agricultural research committees, or CIALs to use their Spanish acronym, are springing up all over Latin America, and they are producing results that are surprising the scientists in the formal research system.

The CIALs bring farmers and researchers together in a process of joint experimentation and learning. The concept was developed at the International Centre for Tropical Agriculture (CIAT) in Colombia, and it quickly caught on. Today there are about 250 active CIALs across Latin America. They vary in size and characteristics, but they all have one thing in common: they provide a direct link between locally organized farmers and the formal agricultural research systems. IDRC has directly and indirectly supported CIALs in Colombia, Ecuador, Honduras, and Nicaragua.

High on the agenda of most CIALs is evaluation of improved local crop varieties, and testing new varieties for suitability in their location. Many of the alternatives tested by a CIAL originate within the local farming community — open-pollinated maize for example. Others, such as hybrids, come from the formal research system. Or there may be a mixture of the two. Management of pests, diseases, soil, water, and nutrients are also significant concerns for the committees. The staple food crops — beans, maize, potato, and cassava — account for more than 80% of the on-farm research that the committees undertake (Table 2).

Once a network of experienced CIALs has formed in an area, the need for intensive coverage by research and extension services is usually greatly reduced because poor rural communities have successfully assumed the task of testing and adapting technology themselves.

The people of the community choose the CIAL's research topic at an open meeting, basing their decision on criteria such as

Table 2. Some key features of 249 CIALs.

Research themes	
Evaluation of crop varieties	62%
Pest and disease management	19%
Soil, water, and nutrient management	12%
Small livestock	5%
Other	2%
Crops researched	
Beans	26%
Maize	16%
Potatoes	14%
Vegetables	13%
Cassava	12%
Fruit	9%
Other	10%
CIALs by gender	
Men only	56%
Women only	7%
Mixed	37%

Source: Ashby et al. (2000).

chances of success, the number and groups of beneficiaries, and the likely costs of the research. Then comes the planning stage, when CIAL and other community members decide on the objectives of the experiment, as well as the treatments and control, the materials and methods to be used, the inputs needed, the data to be collected, and the criteria for evaluating results.

The experiment itself is usually carried out with the help of other members of the community (such as experienced innovators), and once it is completed the CIAL meets with the facilitator (perhaps an agronomist from a local NGO) to evaluate the data collected. In analyzing the results, CIAL members ask "What have we learned?" This stage in the process is especially important when new crops fail or the experiment produces unexpected results.

Finally, the CIAL presents its activities, results, and expenditures at one of the regular open meetings of the community and seeks feedback. The CIAL may also make recommendations based on the results, but it is for the community to decide whether the CIAL

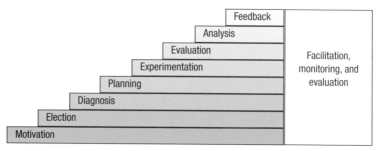

Figure 2. The CIAL process "staircase" (source: Ashby et al 2000).

should continue with the experiment, switch to a new topic, or even cease its activities altogether. The eight steps in the CIAL experimental process can be represented as a staircase (Figure 2).

The Nicaragua experience

That is the CIAL process in theory. In practice the process is usually very dynamic and has both ups and downs. Here's how CIALs came to the mountainous region of Matagalpa, Nicaragua, in 1997. A research team from CIAT, in cooperation with staff from the local Farmer-to-Farmer Program introduced the CIAL to four communities in the Calico River watershed as part of a broader natural resources management research project supported by IDRC and the Swiss Agency for Development and Cooperation (SDC).

In Wibuse, a very poor community in the upper part of the watershed, a committee of men and women experimented with new bean varieties. In El Jicaro, at the midaltitude level, two committees were formed. One with both men and women members experimented with new varieties of maize and beans. The other, composed entirely of women, experimented with vegetables and the use of organic fertilizers. In Piedras Largas, farther down the watershed, the fourth CIAL also experimented with new bean varieties.

There were several failures caused by adverse weather conditions (including a hurricane), plagues, pests, and crop diseases, combined sometimes with inadequate project management. Despite this, the communities evaluated the overall results positively. In Wibuse and El Jicaro, the CIALs set up an additional bean-lines experiment in the second season, assisted by the Central American bean network PROFRIJOL and two national agricultural research organizations. The experiment compared 90 promising lines and one already released variety from Honduras and attracted the attention of many farmers, who had never seen such a diversity of beans. This diversity promised to give them many more production options than in the past. Also, each farmer would now be able to select from a very much larger menu of new materials. The two CIALs, shared responsibility for the management of the plot in Wibuse, and managed to get staff from the national research organization to the field — a novelty in the San Dionisio region.

> Overall, as the Matagalpa experience demonstrates, the strengths of the CIAL system far outweigh any weaknesses.

The following year, members of the four CIALs, together with the CIAT team, organized a watershed-level meeting to share insights, plan for future activities and identify training and technical support needs. Several farmers from other communities who had heard about the CIALs attended the meeting, and subsequently, a number of them took part in the second national CIAL training course. As a result, four new CIALs were formed in the watershed area, and two of the trainees (one man and one woman) became *para-tecnicos*, or junior technical staff. These two assisted the newly operating and existing CIALs, and in 1999 helped to establish two more CIALs in the watershed area. There are currently 14 active CIALs in the San Dionisio Region.

Certainly the CIAL process is not perfect. Most CIALs go through up and down periods resulting from turnover in membership, people's involvement with more immediately rewarding projects, and irregular technical support. Involvement for women is sometimes still difficult. In some CIALs, the efforts and activities remain mostly restricted to the group that can best be called "joiners" — those community-minded individuals who can always be relied upon take part in such activities. But overall, as the Matagalpa experience demonstrates, the strengths of the CIAL system far outweigh any weaknesses.

In Matagalpa, several committees have since moved on to experimenting on a larger scale, addressing new aspects of problems in their communities such as soil fertility. A number of new farmer–leaders have emerged, including several women, and where possible CIALs are linking to each other to exchange ideas and results within the watershed and beyond — through participation in the annual CIAL meetings in Honduras, for example. They also are building bridges to formal research and technology organizations.

Reshaping agriculture in Cuba

Key Research Elements

Cropping system	Beans and maize
Objectives (prioritized)	Diversity / productivity / empowerment
Participation	Researcher-led, collaborative
Social analysis	Gender
Policy component	Increasingly prominent (seed policies)

Necessity drives the search for viable alternatives

Something very like the CIALs is also taking shape in Cuba. Called farmer experimental groups (GICs), they are a key element in a project that aims to reshape agriculture on the island.

The tourism industry notwithstanding, agriculture is still the backbone of Cuba's battered economy. One of the consequences of the economic crisis in Cuba is that agricultural production in the country is moving away from an industrialized, export-oriented, monoculture-based model that is dependent on high inputs. Of necessity, agricultural producers are moving to more diversified, low-input production systems that are oriented to local markets. Another consequence of the crisis has been the rapid deterioration of the conventional, centralized system for seed production, improvement, and distribution.

These unlooked for circumstances have combined to open up a space for agricultural researchers and policymakers to turn their attention to alternative seed production, improvement, and distribution practices, as a crucial contribution to the need to build a new agricultural sector in the country.

In 2000, a multidisciplinary group of dynamic researchers (with backgrounds in biology, agronomy, biochemistry, and sociology) at the National Institute for Agricultural Sciences (INCA) began a project designed to improve the yield and quality of the corn and bean crops through a combined effort of increased varietal diversity and strengthened local farmer organizations. The project is expected to make an important contribution to improving Cuba's food security options.

> Strengthening the organization of farmers increases their capacity to experiment and innovate and to make stronger demands on the formal agricultural research system.

The aim of this innovative project is to strengthen the agricultural biodiversity base in Cuba, making a more diverse and better quality range of varieties available to farmers, agricultural research institutions, and, in the end, consumers. To achieve these aims, the INCA team has several specific goals in mind. First, they want to gain a better understanding of local farmers' knowledge about the management and flow of corn and bean seeds. At the same time they wish to develop a methodology for selecting corn and bean varieties with the involvement of the GICs. Finally, they will disseminate the results obtained by the GICs with the selection, production, and distribution of improved corn and bean seeds.

A secondary but nonetheless important goal is to improve the research capacity of the various agencies involved — including INCA, the GICs, seed companies, and university staff — **through learning by doing**. The project team also believes that strengthening the organization of farmers increases their capacity to experiment and innovate and to make stronger demands on the formal agricultural research system.

Seed fairs and field days: increasing access to diversity

One method the researchers use to introduce farmers to new or unknown varieties or lines is the seed fair. Fairs are organized by breeders and take place at the INCA station. The fairs have proved to be hugely popular, so much so that farmers quite spontaneously started to organize similar fairs in their own communities. Farmers, breeders, and extension agents rub shoulders at the fairs, assessing varieties, and selecting the ones they like best. The materials are then distributed to farmers for testing on-farm. Breeders assist farmers with the experimental design, but all trials are adapted to the local context.

> Results show that women and men farmers have different preferences.

To learn more about farmers' preferences, the project team organizes regular field days, where farmers, both men and women, are interviewed about their preferences. The information gathered is crucial to plant breeders in identifying parental materials and selection criteria. Interestingly, results show that women and men farmers have different preferences. Women farmers select for yield, culinary grain properties, and esthetic features such as colour, shape, and brightness. The men indicate a preference for yield, disease resistance, and pod size. Seeds selected "most liked" are given to the farmers a few weeks after the field day.

Researchers in Cuba have little previous experience with participatory approaches of this kind, so the project team also tends to function as a resource for other researchers who are interested in similar approaches. The team is also involved in genetic analysis through collaboration with biotechnology scientists at the Cuban National Research Institute of Agricultural Sciences.

While Cuba's situation is undoubtedly unique, it is quite possible that a similar collapse of the industrial agricultural sector could occur before too long in other countries of the region, and perhaps beyond. The current agricultural production practices in many countries are highly dependent on expensive technology, and chemical inputs, as well as various kinds of government subsidies, and are simply not sustainable in the long term. **Thus, the Cuban experience will likely have relevance elsewhere in the future.**

CASE STUDIES

Enriching maize and rice in Nepal

Key Research Elements

Cropping system	Rice and maize
Objectives (prioritized)	Diversity / productivity / empowerment
Participation	Farmer-led, collaborative
Social analysis	Gender, ethnicity
Policy component	Increasingly prominent (variety release, IPR)

Local crop diversity and rural livelihoods

For all its Himalayan grandeur, Nepal is only slightly larger than the island of Cuba, occupying less than 0.1% of the world's land surface. Yet its mountains and valleys are home to over 2% of the world's flowering plants. In terms of arable land, the country also has a very high population density — approximately five people per hectare. Most of the people own very small parcels of land, and decreasing fertility combined with land fragmentation (a result of inheritance) have resulted in decreased productivity. In the upland areas the land is marginal at best, and farmers traditionally grow a range of crops in very small plots simply to survive.

It is against this background that a nonprofit NGO called Local Initiatives for Biodiversity, Research and Development (LI-BIRD) works to support sustainable management of renewable natural resources and to improve the livelihoods of people in Nepal. Established in 1995 with its headquarters in the town of Pokhara, 200 km west of Kathmandu, LI-BIRD contributes to conservation and utilization of biodiversity for sustainable development through its participatory research and development initiatives, many of which are supported by IDRC as well as other national and international agencies.

LI-BIRD's activities are wide ranging:

➤ Strengthening the scientific basis of *in situ* conservation of agricultural biodiversity on-farm in different agroecological regions;

➤ Enhancing the contribution of home gardens to on-farm conservation of plant genetic resources to improve the livelihoods of poor farmers;

➤ Supporting participatory crop improvement programs on major cereal crops in high-yield potential production systems;

➤ Developing and refining tools and techniques designed to create awareness at the grassroots level — from diversity fairs and traveling seminars to drama and folk song competitions; and

➤ Conducting policy research on topics such as a seed regulatory framework, government extension and credit policies, agrobiodiversity policies, and land use management to support informed decision-making by the country's policymakers.

Nepal is rich in the diversity of both cultivated and wild relatives of rice, and is home to as many as 2 000 different landraces. Different rice varieties are grown for different purposes — such as home use, festivals, selling, honouring guests, and even medicine. Not surprisingly then, many of the projects undertaken by LI-BIRD and supported by IDRC have been concerned with the improvement of rice production through PPB.

Even with no formal dissemination system, crop varieties can spread over long distances, mainly through the personal contacts and networks.

As far back as 1985, current LI-BIRD staff (then employed by the Lumle Agricultural Research Centre) pioneered PPB work through decentralized testing of cold-tolerant rice in the high-altitude mountain village of Chhomrong. Several other participatory rice- and maize-breeding projects followed, carried out in both favourable or higher productivity areas and unfavourable or lower productivity areas. The projects had a mix of goals including productivity increase, biodiversity enhancement, strengthening farmers' breeding skills, and policy changes; as well as specific breeding goals, as outlined in Table 3.

The village of Chhomrong was among the high-altitude communities participating in a project to monitor the spread of rice varieties from PPB programs. Researchers found that even with no formal dissemination system, crop varieties can spread over long distances, mainly through personal contacts and networks. However, this informal system of dissemination is very slow — typically it is 4 years before farmers exchange or sell new seeds outside their own village. The project highlights the need for an effective method to scale out the process in the interest of the whole community.

Table 3. Selected goals and objectives of LI-BIRD PPB projects.

Project	Goals	Objectives
Chaite and main-season rice in favourable production region (Chitwan)	• Productivity increase • Biodiversity enhancement • Research efficiency • Policy change	• Develop varieties for low water regimes • Improve Masuli rice for disease tolerance and yield • Eliminate awns and increase height in Pusa basmati rice • Improve grain quality of IR44595 • Improve CH-45 for disease-tolerance; increased seed dormancy in field
Landrace rice in favourable/ unfavourable region (Jumla, Kaski, Bara)	• Biodiversity enhancement • On-farm conservation • Farmers' capacity building • Policy change	• On-farm conservation through value addition • Improvement for locally important traits
Upland *ghaiya* rice in unfavourable region (Tanahu)	• Productivity increase • Biodiversity enhancement • Users' needs/preferences • Policy change	• Diversity deployment • Drought tolerance
Farmer-led maize in unfavourable region (Gulmi)	• Farmers' capacity building • Users' needs/preferences • Productivity increase • Biodiversity enhancement • On-farm conservation	• Address lodging problem of Thulo piyanlo landrace • Diversity deployment
High-altitude rice in unfavourable region (Maramche, Kaski)	• Farmers' capacity building • Users' needs/preferences • Productivity increase	• Address shattering problem (Machhepuchhre-3 rice) • Develop cold-tolerant, farmer-accepted variety

Source: Adapted from Subedi et al. (2001, pp. 75–86).

Another project studied upland rice, locally known as *ghaiya*, which is grown under rainfed conditions on flat land, terraces, or hill slopes of newly cleared forests. This crop is mainly grown by poor farmers on unirrigated ancient alluvial river fans call "tars." *Ghaiya* has considerable importance in the farming system and is preferred to maize both for its food value and its straw for animal feed.

The study found that *ghaiya*-growing farmers possess a wealth of knowledge in managing their soils to maximize the crop yield. Farmers also demonstrated that mixed cropping *ghaiya* with maize results in a higher combined yield, and has a practical advantage in that a few rows of maize in the flat tars make it easier to broadcast *ghaiya* more uniformly. Some farmers, however, prefer to plant maize after *ghaiya*, saying this helps to maintain soil fertility. Diversity still exists for indigenous *ghaiya* varieties, although the number of varieties farmers maintained varies according to the size of their land holding — the larger the holding, the more varieties. In a majority of the areas studied, farmers maintain at least two varieties with different maturity.

In the Pokhara Valley, where rice is grown both as a staple and as a cash crop, another IDRC-supported project studied the rice-growing environments and status of indigenous rainfed and aromatic rice varieties. Farmers here reported more than 75 local landraces, yet only 11 of these were widely cultivated. Seeds of all the landraces have been collected for conservation, study, and possible promotion.

Insights and achievements

Over the years, LI-BIRD's pioneering work in Nepal's various agroecological regions has resulted in important achievements and insights:

→ The direct participation of farmers often leads to new breeding objectives; therefore, the methodology should be cyclical and adaptive, not linear and rigid. Biodiversity seed fairs, kits, and community registers are useful tools to encourage farmers' involvement.

→ Involving farmers in the planning process results in breeding objectives that are much closer to farmers' perceived needs and interests.

- Within larger, high-productivity regions there exist diverse niche environments associated with different user preferences; different options are needed to address this biophysical and social variety.

- PPB has the potential to increase diversity. It accelerates change by introducing genes and genotypes as key inputs in the ongoing process of *in situ* conservation of crops.

- There is no blueprint for the right form and amount of participation by women and men farmers in the process. However, defining and agreeing on clear responsibilities is essential to manage or guide the process.

Maize improvement in southwest China

Key Research Elements

Cropping system	Maize
Objectives (prioritized)	Diversity / productivity / empowerment
Participation	Researcher-led, collaborative
Social analysis	Gender
Policy component	Research policies, plant breeding, seed policies

Bridging the worlds of farmers and scientists

In the remote and harsh uplands of southwest China, farmers must struggle to eke out a living, unlike those fortunate enough to farm in the Northern Plain — China's "corn belt." Yet this remote region is one of the places in the world where people first began to cultivate maize. Farmers here have cultivated and relied on maize for their survival for countless generations, and they maintain a higher level of maize varietal and genetic diversity than in the rest of the country. Today this region is a treasure trove of maize genetic diversity that is vital to the future of maize cultivation in China.

Maize is now the most important feed crop and the third most important food crop in China. It is the main staple food crop for the rural poor in the upland areas in the southwest. The government of China has followed a modern technology-oriented approach, relying predominantly on its formal seed system. The development and distribution of modern varieties, mainly hybrids, for the three main staples — rice, wheat, and maize — has been the core task and the first priority for the formal system to achieve the overall goal of national food security.

Hybrid maize is now grown on approximately 80% of the total maize-production area in China, particularly the uniform and high-potential areas of the Northern Plain. The introduction of a market economy has resulted in an increasingly profit-driven seed production and supply system. Hybrid breeding and hybrid seed production have drawn more attention and investments than ever before. Conversely, a study done in one of the southwest provinces, Guangxi, revealed that more than 80% of the seed supply is from farmers' own seed systems, maintaining diversity for the interests and sustainable livelihoods of all farmers.

The genetic base for maize breeding in China has been dramatically reduced during the last decade. Although the total national maize germplasm collection has around 16 000 entries, 5 dominant hybrid maize varieties now cover 53% of the total maize-growing area in the country. In Guangxi province the total maize germplasm collection has around 2 700 entries; among them, more than 1 700 are landraces from the region. However, the utilization of these collected materials in breeding is very limited. Only 3 main hybrid breeding crosses are used and all the 14 hybrids bred out in the last 20 years share the same inbred line to different degrees. Meanwhile, in several provinces landraces in farmers' fields are degrading and disappearing as a result of the continuing spread of modern varieties.

Although China's economic growth has been impressive, poverty remains persistent in many rural areas including Guangxi province — in particular affecting women and households headed by women. Rapid growth also goes hand-in-hand with increasing natural resource degradation. Change at the political level comes at a slower pace. Top-down planning and decision-making are still the norm at various levels of government. But even here spaces can be seen to be opening up.

A cooperative and complementary relationship between the two systems is urgently needed to address the challenges in food security and biodiversity facing China today.

This is the background to a research project begun in 1999 by the Center for Chinese Agricultural Policy (CCAP) in collaboration with the Guangxi Maize Research Institute (GMRI). The project built on a study financed by the International Maize and Wheat Improvement Center (CIMMYT). This study was carried out by a Chinese doctoral student, Yiching Song. In assessing the impact of CIMMYT's maize germplasm on poor farmers in southwest China, she looked in particular at the processes of technology development and diffusion by both the formal and the informal systems.

One of the study's main conclusions is that a cooperative and complementary relationship between the two systems, rather than the current separated and conflicting situation, is urgently needed to address the challenges in food security and biodiversity facing China.

The research project, supported by IDRC and the Ford Foundation, set out to identify and assess ways of developing a mutually beneficial partnership between the formal and informal systems in maize crop development specific to the southwest region. Two key goals:

➤ to better promote and use the techniques that enable indigenous local communities to conserve biodiversity and

➤ to find ways to involve those communities in the design and implementation of on-farm biodiversity conservation.

The project team members come from several institutions and groups. They have different disciplinary backgrounds and operate at different levels. Five women farmer groups, six villages, six township extension stations, two formal breeding institutes, and CCAP have been directly involved in both the design of the project and its implementation. Now in its second phase, the project is attempting to link community-based action research with the policy-making process by increasing efforts to engage directly key decision-makers in the maize policy arena at both the provincial and national levels.

The field experiments have proved to be effective in strengthening interaction, communication, and collaboration among the stakeholders.

The field experiments use both a researcher-led and a farmer-led approach with different research focuses in each trial for comparison. More than 40 varieties were identified as target varieties for PPB and PVS trials at the GMRI station and in five villages. From the 40 trial varieties, 3 preferred varieties were selected by farmers for agronomic, cultural, and economic reasons. These varieties have been released and used in the trial villages and in neighbouring villages. In addition, five exotic varieties from CIMMYT have been locally adapted, and five landraces from the trial villages have been improved through the joint efforts of farmers and breeders. An improved variety (in terms of local adaptation and farmer preferences) from women farmers has been tested and certified by the formal breeding institution and is widely used in the project region. Formal breeders have identified some very useful breeding materials and inbred lines that have a very broad genetic base from the landraces in farmers' fields.

There are other benefits. The field experiments have proved to be effective in strengthening interaction, communication, and collaboration among the stakeholders. They have also strengthened the local-level organizational and decision-making capacity of farmers. And among the formal breeders there has been impressive change in attitude — the needs and interests of farmers are now considered and included in the breeding plan and research priorities of the institutions, and farmers' efforts and knowledge in genetic biodiversity management are increasingly recognized by policymakers at both provincial and national levels.

The success of the project has led GMRI to combine gene bank conservation with *in situ* conservation of landraces. In addition, the China Crop Science Institute will include the local germplasm conservation efforts in Guangxi in its national plan for broadening the genetic base. **Meanwhile, CCAP has played a crucial role in expanding the impact and influence of the results at national policy levels.** For example, the project was presented and discussed at a national policy-planning workshop coordinated by CCAP and CIMMYT in Beijing in March 2002. This important conference was the first time that 40 prominent national agricultural policymakers and maize researchers had discussed the participatory approach as an alternative and complementary methodology for crop improvement and agrobiodiversity management.

The global program on Participatory Research and Gender Analysis

Key Research Elements

Cropping system	Open-pollinated, cross-pollinated, vegetatively propagated crops
Objectives (prioritized)	Productivity / empowerment / diversity
Participation	Variable, consultative, and collaborative
Social analysis	Variable, gender
Policy component	Variable, IPR, seed policies, variety release, research policies

Emphasizing women's roles

Perhaps the most extensive program in support of PPB on a global scale is sponsored by the CGIAR. Called the Program on Participatory Research and Gender Analysis (PRGA), its goal is to "assess and develop methodologies and organizational innovations for gender-sensitive participatory research and to operationalize their use in plant breeding and in crop and natural resource management."

The PRGA is cosponsored by four of the IARCs, and its activities are funded by national governments and several donor institutions, including IDRC. The program's members include NARS, NGOs, and universities around the world. As the name indicates, the PRGA places considerable emphasis on the roles of rural women in managing plant genetic resources.

> Women throughout the developing world have detailed knowledge of, and strong preferences for, specific crop traits.

The emphasis on women's roles and needs is a logical outgrowth of 20 years of effort to make science more responsive to poor farmers. Women play many roles — growing, harvesting, storing, and preparing food. Perhaps none is more important than their role in plant breeding. Women farmers are prolific and adept plant breeders, as well as key managers of natural resources such as soil and water. They domesticate wild species and play a vital part in selecting and storing seeds for future crops. **Women throughout the developing world have detailed knowledge of, and strong preferences for, specific crop traits, and studies show that men and women often have markedly different expectations and knowledge of crops — differences that research and policies need to take into account.**

Projects under the global PRGA program support the worldwide development and assessment of gender-sensitive participatory research methods. The goal is to introduce proven approaches into the IARCs and eventually into national programs. Several of the activities already described in this chapter come within the scope of the PRGA. For example, the CIALs in Latin America, the ICARDA barley research in the Middle East, and LI-BIRD's upland research in Nepal. The research teams in China and Cuba have also established links with the PRGA program.

One of the program's key strategies for advancing gender-sensitive PPB is a competitive small grants program. In Peru, for example, a grant enabled women to be involved in selecting new potato clones, giving them greater decision-making power and control over resources. As has been found in other regions, the women's choices were different from those of the men. Participatory approaches applied in Uganda have resulted in men working more with women; in Kenya, they led to increases in the number of women in local management committees.

Although the small grant projects are the PRGA's main arm in the field, the program's staff are also engaged directly in cutting-edge research. For example, conducting a study that addresses the challenging issue of how to attribute intellectual property rights that emerge from collaboration between researchers and farming communities. This work starts to fill a major gap in the international arena, where current agreements draw prime attention to the rights of plant breeders and farmers, but fail to address the division of benefits that could result from collaborative work.

The benefits of participatory research have been documented, but to persuade more scientists and research managers to begin to incorporate these approaches into their research, it is vital to be able to compare the participatory approach to other, more traditional approaches. Program staff have developed and applied tools for empirical impact studies in both PPB and natural resource management. Both impacts and costs were studied, with a particular focus on documenting process impacts of different types of participatory research, as well as the impact of involving farmers at different stages of research.

Initial findings suggest that involving farmers more closely in the research process and giving them more control yields many positive impacts, including increased profits for the farmers. There is also empirical evidence that participatory research reduces costs by helping to prevent the development of technologies that are not subsequently adopted by the intended users. For example, feedback from Indonesian farmers at an early stage of sweet potato research led researchers to modify their proposed technology.

> Involving farmers more closely in the research process and giving them more control yields many positive impacts, including increased profits for the farmers.

In a further effort to promote and facilitate the use of participatory approaches, the PRGA has built a network of knowledge and practice including NGOs, NARS, and IARCs. E-mail lists encourage an ongoing worldwide exchange of expertise, while international seminars bring together hundreds of practitioners from around the world. PRGA staff have created three publicly accessible databases with information on projects that use gender-sensitive participatory approaches and have established a network of PRGA liaisons and gender focal points in all the CGIAR centres. In addition, staff have organized and participated in numerous training workshops on participatory research and gender analysis methods and have published several training manuals.

What has been accomplished?

Table 4 summarizes some of the key results obtained in the six projects described above. In Part 4 we will see how these projects and their results fit within the overall scope of IDRC's biodiversity research program.

Table 4. Selected key results of sample projects.

	Participatory barley improvement (ICARDA)	Local agricultural research committies (CIALS)	Reshaping agriculture (Cuba)	Enriching maize and rice (LI-BIRD, Nepal)	Maize improvement (China)	Participatory Research and Gender Analysis (PRGA)
Methodology innovation	PVS and PPB methods pioneered in Middle East and North Africa; decentralized trials; major attention to genotype–environment interactions	Community-based, farmer-led, on-farm, experimentation method developed in Latin America; a new form of local, rural development organization	PVS and PPB methods pioneered in Cuba; introduction of (diversity) seed fairs; introduction of CIAL-like methodology	PPB and PVS pioneers in Nepal; farmer research groups established; seed fairs popularized	Supporting farmer-led breeding efforts, strengthened local organization; small (seed) enterprise development	Supporting and moving PPB and PVS forward in the CGIAR, NGOs, and NARS; state-of-the-art global overviews documenting and analyzing good practices of NGOs, NARS, and IARCs
New partnerships	CGIAR–NARS collaboration in various countries established	Farmer groups connected to NARS, NGO, CGIAR centre; extension staff fully involved in research	NARS–cooperatives and farmer groups	NGO–NARS, NGO–CGIAR collaboration	NARS–extension bureau–farmer groups	CGIAR–NARS–NGO partnerships
Breeding results	Improved varieties, diversity increased; farmers' effort recognized at national level	Diversity increased, improved varieties	Diversity increased, improved variety (maize)	Improved varieties, diversity increased	Improved varieties; farmers' effort recognized at provincial level	Diversity increased, improved varieties

Improved varieties and (or) increased diversity lead to improvements in food access, diversity, quality, amount, or a combination of these. Examples: Early maturing varieties fill a gap in food-deficit months of the year. Higher yielding varieties contribute to putting more food on the table. More robust varieties require fewer inputs such as fertilizer and (or) water or are more resistant to plagues and diseases. Varieties with improved grain quality enhance nutrition.

Capacity building	Attitudes and skills of scientists changed and strengthened; recognition of farmers' knowledge and capacities; farmers' research skills improved	Farmers' research and organization skills strengthened; attitudes of scientists changed; women recognized as skillful experimenters and breeders	Attitudes and skills of scientists changed and strengthened; farmers' research and organization skills strengthened; women recognized for their experimenting and seed-selection skills; women empowered to discuss research results and to explore seed production as a source of income	Farmers' research and experimentation skills strengthened	Attitudes and skills of scientists changed and strengthened; farmers' research and organization skills improved; women breeders recognized for their knowledge and expertise	Researchers' attitudes and skills changed; farmers' research skills strengthened; women's key roles in agrobiodiversity management highlighted
Policy changes	PPB accepted by NARS; decentralization of breeding trials now a common feature	Methodology accepted by NARS (Colombia)	Participatory approach gaining ground in research cycles; application in other crop-improvement programs (rice, tomato)	Farmer-bred varieties officially released	Farmer-bred variety released; participatory approach gaining ground in research cycles	PPB endorsed as organic component of breeding research in CGIAR

Part 4

Learning from Experience

Participatory plant breeding

Part of the solution

The preceding chapter contains six stories — a cross-section of a decade of research in countries around the world, some of which is still ongoing. These stories show us farmers, even whole communities, becoming empowered, gaining a small measure of control over their natural resources. They show us researchers and scientists discovering the value of extending their work outside of the laboratory and research station to benefit from the knowledge and experience of the men and women who are closest to the land. Of course, any good story contains a message — a lesson — and these stories are no exception. What lessons can be learned from examining 10 years of research? And how can those lessons be used to enhance decision-making about future directions for agricultural and environmental policy and research?

The issues are clear — genetic erosion is taking place at an alarming rate. We have seen the example of China, where the total national maize germplasm collection has around 16 000 entries, but 53% of the country's maize-growing area is planted to just 5 dominant hybrid varieties. This pattern is repeated in most of the world's food-producing areas, and brings an entirely new meaning to the expression "doing more with less."

The other side of the coin: the world's poor depend on biological products for as much as 90% of their needs — food, fuel, medicine, shelter, transportation. Approximately 1.4 billion people, mostly poor farmers, have traditionally used and improved their own crop seeds, leading to the development of many landraces for each crop. About 75% of the world's population relies on traditional medicines for primary health care. These biological resources, including the skills and knowledge that have contributed to this diversity, are increasingly threatened by changes in social, economic, and political structures, changes in natural environments, and increasing pressure on the resources themselves.

The threat to biodiversity is a complex and broad-ranging issue that ultimately affects all of us — South and North — because it has the potential to disrupt our food supply. If PPB is at least a part of the solution, then it is important to understand what has been accomplished by this relatively new research approach, and how those results might be applied most effectively on the wider stage.

Recall that the approach to agrobiodiversity research outlined in Part 2 of this book is based on three widely accepted objectives: enhancing knowledge about agrobiodiversity, designing and testing practices and measures that add value to agrobiodiversity, and creating supportive policy and legislative alternatives. The search for lessons begins here, but underlying these objectives we also need to look for more general outcomes that might point the way to future research and policy development (see Table 4). For example:

➤ research, advocacy, and policy methodology development and innovation,

➤ individual and organizational research capacity building,

➤ new or stronger partnerships created among stakeholders or across sectors, and

➤ clear research and policy responsiveness to user needs.

Finally it is important to determine if there have been any direct or indirect spin-offs from the research. Has the research been adopted and adapted successfully elsewhere? Has there been a "multiplier effect"? Has the research led to new approaches, new thinking? In short, have the lessons been not just learned, but also applied? Have we learned from our experiences?

Agrobiodiversity

Crucial to document and integral to people's lives

In the effort to enhance existing knowledge about agrobiodiversity, researchers have dedicated much time and effort to documenting and describing existing crops and cropping systems, as well as local or indigenous knowledge about these. Their findings suggest that in many places **the maintenance of agrobiodiversity is integral to people's cultural identities**, and is usually a response to environmental, ecological, and economic uncertainty and fragility. Sometimes it is simply a way of making the best use of niche conditions.

Diversity and farmers' knowledge about the dynamics of that diversity is alive and well in many places, although as we have seen there is mounting evidence that traditional cropping systems and crops are under increasing pressure. These cropping systems include grains, root crops, legumes, spices, forages, and so called "uncultivated" or "wild" foods.

> There are numerous examples of research tools "traveling" from one site to another — from researcher to researcher, farmer to farmer.

The pressure on these systems comes in part from market forces, but a number of other factors come into play. Science threatens traditional crops and cropping systems through the introduction of a limited number of hybrid varieties and the subsequent replacement of more diverse mixes of traditional varieties. Pressure also comes from human activities, such as migration, "modernization," and,in some areas, warfare. Natural disasters such as hurricanes, floods, and earthquakes are an ever-present and unpredictable factor. Resource tenure also has emerged as a key factor in relation to the dynamics of diversity, and the space farmers have to maintain or increase varieties. More detailed research is needed to gain a better understanding of these factors and to identify possible entry points for action.

The dynamics and trends of agrobiodiversity are now much better documented and analyzed. Although the depth and quality of the results vary, several innovative tools have been developed that enable detailed analysis of trends at various levels — crop, field, and farming systems. Significantly, there are numerous examples of these tools "traveling" from one site to another — from researcher to researcher, farmer to farmer.

A good deal of valuable knowledge has been gained about the interrelations between the human factors and biophysical factors at various levels. The human factors include knowledge, skills, needs, and interests expressed through gender, age, class, and ethnicity. Documenting how they affect outcomes in combination with the biophysical factors at crop, cropping system, and landscape levels offers important insights into how and why communities conserve biodiversity.

One of the earliest biodiversity projects supported by IDRC was a study begun in 1992 that examined the factors maintaining sorghum landrace diversity in Ethiopia, which is thought to be the original home of this important small grain. This project has since been expanded and is ongoing. It has documented — through observations, surveys, and interviews — the vast taxonomical knowledge of Ethiopian farmers and confirmed their role in the maintenance of sorghum landrace diversity in the north Shewa and south Welo regions, as a means to reduce the risk of homogenization.

In addition, the study documented farmers' knowledge about storage conditions and duration of sorghum landraces, and the action needed to reduce losses caused by pests. The research focused on two areas: the dynamics and trends of crop diversity; and farmers' selection criteria at the field, community, and agro-ecosystem levels, from a gender perspective. It also examined the many variables that influence the use and management of diversity.

Others have studied complex and highly diverse cropping systems such as home gardens and slash-and-burn farming systems in Central and South America, as well as indigenous vegetable gardens in East and Southern Africa. A few have studied inte-grated crop-and-animal systems, such as the diverse fish–rice systems in the Mekong Delta region in Viet Nam.

Women all over the world play key roles in the management of agrobiodiversity. Recognition of these roles, however, remains weak in many places. There is also a need for more systematic and rigorous study of the diverse roles of women and men, and of the gender-differentiated impacts of changes in diversity. Meaningful research in this area requires going beyond simply breaking down data by the sex of participants. For example, age, ethnicity, economic status, and level of education are all significant factors that need to be considered.

There now exist a wide array of innovative research methods. One of these is the small grants mechanism that provides support for a wide variety of individual projects and methodologies under the "umbrella" of a common research theme. Another method is the multicountry research program. This provides a common framework for a shared research approach and methodology and for networking functions.

Successful spin-offs

Adding value to the conservation and use of agrobiodiversity

In most PPB-focused research, adding economic, social, or cultural value has been a core element of the process from the beginning; in others, it is initiated only following an extensive period of documentation and description. Much depends on the project design and approach, but in either case the results have been impressive.

Research tools and methodologies that "travel" from one region, or country, to another represent another way to add value to the original research investment. Therefore, achieving direct or indirect spin-offs to maximize the impact of research results is a goal of many of the projects.

As we have seen, the successful CIAL methodology developed by CIAT in Colombia has spread throughout much of Latin America. Approximately 250 farmer groups are carrying out experiments to increase crop diversity, improve productivity, and conserve soil and water. Most recently, elements of the CIAL approach have also made the move to China and to Cuba. Similarly, the maize project in southwestern China included elements of a maize-improvement project supported by IDRC and the CGIAR in Mexico. More examples: in Nepal, farmers in villages neighboring the research site picked up elements of the research and started their own experiments. And following successful work in Morocco, Syria, and Tunisia, ICARDA is building on experience gained and lessons learned in the barley-improvement project in North Africa and the Middle East in other countries in the region and beyond.

Some of the research has gained international recognition. For example, the Community Biodiversity Development and Conservation (CBDC) program was recommended by the FAO in its 1998 report on the *State of the World's Plant Genetic Resources for Food and Agriculture* as a model for *in situ* conservation approaches. The CBDC is an interregional initiative developed by several NGOs and government agencies in Africa, Asia, Latin America, the Netherlands, Norway, and Canada.

In 1996 IDRC helped to organize two international workshops, one in India and another in the Netherlands, which proved to be catalysts in developing new and innovative collaborative efforts in the field of crop biodiversity. At the India workshop, plant breeders, scientists, and policy activists assessed thinking and practice in South Asia about agricultural biodiversity. They quickly recognized that they shared a number of ideas and interests, and explored convergence among perspectives as well as avenues for future collaboration.

One concrete outcome of the India workshop was the creation of the Using Agricultural Diversity small grants program in 1997. These awards help grassroots organizations and scientists working with farmers in South Asia to undertake applied research on the use of agricultural diversity — including wild herbs, crops, and livestock — to meet the needs of farm households and to protect the environment. The awards also encourage research collaboration, exchanges, and dissemination of information among the formal and informal sectors on practical means to enhance the sustainable use of agricultural diversity.

The workshop in the Netherlands brought together a dynamic group of like-minded researchers and staff from the CGIAR system, the FAO, European government agencies, a number of NARS, and donor agencies to explore common issues, interests, and methodologies. The participants took stock of PPB efforts deployed by plant breeders, conservationists, and social scientists, and developed concepts for cooperation to stimulate further research and practice in decentralized PPB.

More added value: this workshop also planted the seed for what is now the PRGA. The PRGA was first constituted unofficially in 1996 and subsequently formalized as a CGIAR program the following year.

> There is definitely a need for a more systematic approach to a whole range of trade-related issues.

Adding value has also been achieved by linking *in situ* and *ex situ* conservation, and by strengthening or improving seed-production systems through mechanisms such as seed fairs and seed banks. These elements are part of the INCA project in Cuba and the maize-improvement project in China. Also, a project in Costa Rica has attempted to broker options for the commercialization of tapado beans as certified organic produce, both in the country and abroad — a good example of trying to add value from the demand side. There is definitely a need for a more systematic approach to look at, and deal with, not only organically certified produce but also a whole range of trade-related issues such as unfair trading practices and market linkages.

Much knowledge has been gained and many skills acquired through **learning by doing**, as well as through projects designed specifically with training in mind. The research has contributed significantly to strengthening individual and organizational research, documentation, and management skills. The list of leading PPB-minded organizations includes a wide range of research-oriented bodies, among them NGOs such as LI-BIRD and the Southeast Asia Regional Initiatives for Community Empowerment (SEARICE, an NGO that is a member of the CBDC program); NARS such as INCA in Cuba and GMRI in China; universities from Can Tho University in Viet Nam to the universities of Guelph and Ottawa in Canada; and international centres such as CIAT, CIMMYT, ICARDA, the International Crops Research Institute for the Semi-Arid Tropics (ICRISAT), and the International Plant Genetic Resources Institute (IPGRI).

Many of the projects carried out by these organizations represent truly pioneering endeavours — and were often risky because the researchers dared to undertake institutionally novel and methodologically innovative research, rowing against the stream of conventional breeding practice. Success was not assured, and many of their colleagues were skeptical about the prospects of PPB. Two good examples of risk-taking that are paying off handsomely are the INCA project in Cuba and the maize-improvement project in China. Both involve NARS and both are carried out in countries where participatory approaches are simply unheard of. The work by ICARDA on barley in North Africa and the Middle East is another example of successful risk-taking.

The PRGA program and its members have made some important strides toward changing research policies within the CGIAR itself. An example of a crucial step forward is the recommendation made in 2000 by an advisory group to the CGIAR Technical Advisory Committee (TAC) that PPB become an integral part of each CGIAR centre's plant-breeding program. Another sign that the innovative work underway is getting recognition was shown when the ICARDA barley research team headed by Salvatore Ceccarelli won a prestigious CGIAR publication award in 2000. More profound changes, however, still proceed slowly.

Through efforts to collect and synthesize research issues, project approaches, and methodologies at the global level, the PRGA program has produced a number of important, comprehensive reports. These reports discuss the technical and institutional issues in PPB from the perspective of both formal plant breeding and farmer plant breeding. The insights contained in these reports capture the cumulative experiences of many teams and projects. In 1999 the program's Plant Breeding Working Group also produced detailed guidelines for developing PPB projects.

Other projects have produced extensive training materials for specific methodologies, such as CIAT's series for use by the CIALs in Latin America. Based on its global *in situ* program, IPGRI has produced a training guide for on-farm conservation of crop genetic diversity.

> The project experiences make it clear that farmers can and should play a key role in the research process through PPB.

What works best, when and where? It is still early to determine the full and long-term impact on farmers' livelihoods and the viability of the numerous approaches that have been and are being supported. However, diversity is gaining terrain in many places. Yield results are improving in quantity, quality, or both. Capacities have been strengthened. It is also clear that there is no one-size-fits-all solution to agrobiodiversity conservation and crop improvement — diversity in methodology is needed to mirror diversity in rural contexts.

The experiences make it equally clear that farmers can and should play a key role in the research process through PPB, and that new cultivation options can and should be introduced in a participatory manner. These findings can be singled out as positive results. On the negative side of the ledger, an important constraint identified by several projects has been policy related — specifically, the lack of a supportive policy environment. There is generally a lack of policies that support change.

Seeking policy and legal alternatives

National and international support for local efforts

Government policies — or the lack of them — can and do have an enormous impact on efforts to conserve agrobiodiversity in all parts of the world. In many countries the issue is not even considered by government agricultural officials. In others it is often seen as a side issue, not one that is central to the long-term sustainability of the food supply.

Several attempts to better understand and perhaps modify these attitudes have been made by studying how current policies (or the absence of policies) influence agricultural biodiversity. As a result, policy and legislative alternatives in the form of incentives, regulations, or legislation have been proposed. For example, the PRGA program has conducted policy research around questions of recognition of farmers' contribution to agrobiodiversity conservation and crop improvement, as well as intellectual property rights.

It has proven difficult to link local-level users' perspectives and interests with macro-level policies, and to do so in an inclusive manner. For example, how do we ensure that all stakeholders have a say, or that there are better links between researchers and extension agents, or between researchers and policymakers? It is also clear that longer term studies are needed to monitor policy impact, such as an analysis of policy disincentives. Integration of analyses at different levels is complicated, and it is important that research on policy and legislation be integrated into research that focuses on cropping systems and knowledge enhancement.

> There is not yet a very strong critical mass of researchers willing to get involved with policy-making and implementation debates and processes.

It is also essential that more researchers begin to take an active interest in the impact of policies and legislation on biodiversity in general and PPB in particular. Although they are growing in number and strength there is not yet a very strong critical mass of researchers involved – or willing to get involved – with policy-making and implementation debates and processes.

In Zimbabwe, there is a national project to develop specific legislation on IPR patents. This project has produced significant inputs for the national policy and legislation design and drafting processes. A similar project is underway in Viet Nam. The Zimbabwe project also has a regional impact as it is part of an effort by the Organization of African Unity (OAU) to draft model legislation for its member countries. Also at the regional level is a study of plant breeders' rights in Latin America.

At the global level the work of the Crucible Group on a series of issues dealing with rights, laws, and legislation has been innovative and enlightening. The Crucible Group is a global think-tank that brings together 45 individuals from 25 countries – from industry, government, advocacy groups, and aboriginal groups. All serve in a private capacity and share a common interest in preserving and developing the world's genetic heritage. Their role is to critically discuss issues and formulate policy recommendations related to the use, conservation, and ownership of plant genetic resources. Results of these discussions – that encompassed both consensus and disagreements – were first published in 1994 in the IDRC book *People, plants, and patents: The impact of intellectual property on trade, plant biodiversity, and rural society.* The book contains 28 recommendations.

Following a new series of discussions, insights and recommendations were updated and jointly published by IDRC, IPGRI, and the Dag Hammarskjöld Foundation in 2000 as *Seeding solutions. Volume 1: Policy options for genetic resources.* This volume included 15 new recommendations. Volume 2 of *Seeding solutions* was published in 2001 and outlines legislative options for national governments, for the conservation and exchange of germplasm, the protection of indigenous and local knowledge, and the continued promotion of biological innovations. More recently, IDRC and several other donors, in collaboration with IPGRI, created the Genetic Resources Policy Initiative (GRPI) to build on and expand the policy work of the Crucible Group.

Other studies undertaken as part of crop-related projects have dealt with issues related to international policy-making bodies such as the CBD and the World Trade Organization (WTO). These efforts have contributed to raising broader awareness, putting issues of importance on the agendas, and bringing the voices of indigenous peoples to the negotiating table.

Summing up

What has been achieved?
Awareness and understanding of the importance of agrobiodiversity have increased considerably during the last decade. There now exists a wealth of knowledge about local plant genetic resources and resource uses — often in remote places that are hard to access. This knowledge covers the three cropping types, three continents, numerous agroecozones, and a variety of indigenous systems.

The cumulative research results constitute a sound body of research that has contributed to put agrobiodiversity on the international research agenda as a topic of broad interest — including the agendas of donor organizations that fund international development or development research.

Specifically, a series of new, interdisciplinary methodologies is available to study agrobiodiversity and to strengthen local capacities to maintain and increase diversity. These methodologies combine consultative and collaborative participatory elements, on-farm and on-station experimentation, and, to varying degrees, a user-differentiated analysis based primarily on gender.

Methodologies and tools have been documented and made accessible to those interested in studying or implementing agrobiodiversity dynamics.

In some countries and organizations, teams continue to pioneer these new methodologies. Wherever this happens, the projects are truly showcases, often drawing considerable attention and sometimes critical scrutiny as well. The institutionalization of methodologies and approaches has been put explicitly on the agenda of a few initiatives, most notably the PRGA program of the CGIAR. Some innovators have already succeeded in integrating such methodologies into their national research agenda: for example, CIMMYT in Mexico or ICARDA in Morocco and Tunisia. Several other project teams — including those in China, Cuba, Ethiopia, Jordan, and Nepal — aim to influence national research agendas in the near future. The Crucible Group, through its insightful publications, has been instrumental in getting agrobiodiversity issues (such as indigenous knowledge, farmers' rights, and access and benefit sharing) onto national and international policy agendas. Alternative regulations, agreements, and model laws have been formulated, proposed, and advocated. However, as this review clearly indicates, much work remains to be done to strengthen and expand the field of agrobiodiversity conservation, crop improvement, and PPB.

Proof of progress

In late 2002, the PRGA in collaboration with another CGIAR initiative, the Systemwide Genetic Resources Programme (SGRP), organized a workshop entitled "The quality of science in participatory plant breeding." The workshop brought together 35 PPB research and research-management innovators from the around the world — representing the CGIAR, NARS, NGOs, and donor organizations. Their goal was to take stock of and debate future concepts and methodologies applicable to PPB. The topics they debated included priority setting, on-farm testing and evaluation, scaling up, measuring impact, putting PPB in a more holistic context, and PPB and biotechnology.

> These partnerships are not only changing research practices, they are also laying the groundwork for future innovation and longer term results.

The 5-day workshop was clear proof that impressive progress has been made in the PPB field. Concepts and methodology have deepened and matured; the number of practitioners and organizations supporting PPB work has increased and expanded; a start has been made to link PPB work more closely with integrated natural resource management, with seeds systems, and with access, recognition, and benefits issues.

The workshop strengthened existing partnerships and forged a number of new ones — a feature that sometimes goes unnoticed in assessing the value of such events. These partnerships are not only changing research practices, they are also laying the groundwork for future innovation and longer term results toward more user-responsive research.

Part 5

Recommendations

Participatory plant breeding demands a different, innovative way of addressing human needs that goes well beyond the aim of increasing productivity. Its goals are achieving productivity increase, diversity enhancement, and empowerment.

As we have seen, dynamic approaches that are collaborative, involve multiple stakeholders, and employ sound participatory methods do contribute to food security and improved livelihoods. However, field-level interventions alone, both on the farm and in communities, are not enough to sustain these well-tested alternatives. **Long-term success requires that these efforts be backed up by supportive policies, by actions to ensure that policies are implemented, and where necessary by related legislation.**

Bridging the divide between research in the field and widespread implementation of the methodology needed to support the processes that maintain diversity over time represents a major political challenge. Meeting that challenge requires affirmative action in six decision-making areas of government or research, or both. Relevant policies concern agricultural development, conservation of natural resources, variety release, intellectual property rights, farmers' and plant breeders' rights, marketing, and product certification.

1. Increase relevance: putting users of crop diversity at centre stage

PPB is particularly relevant in any of the following situations:

In marginal, stress-prone regions	When improvements are desired in species of vital interest to farmers that are not served by research otherwise, such as minor crops (often women's crops)
In agroecologically or socio-culturally heterogeneous regions	When farmers cannot sustain breeding efforts without assistance, such as when they are faced with diseases that require multiple and specific new traits
When productivity and diversity are twin goals	When farmers want to gain greater control over crop breeding and seed production
When diversity maintenance and enhancement is the main goal	When crop improvement efforts have become ineffective or even counterproductive as a result of discord among various stakeholders

Attempting to address any or all of these challenges requires a willingness to work with a new research and development model: an approach that contributes to increasing both effectiveness and efficiency by putting the users of crop diversity at centre stage.

2. Create new partnerships: farmers, scientists, and other stakeholders working together as equals

A new division of labour, new partnerships, and new forms of cooperation are at the heart of the PPB approach. The goal must be to involve farmers in the research in ways that are meaningful to them: in short, to improve the quality of farmer participation. Farmers must no longer be just the passive recipients of technologies, seeds, and information. They must be encouraged to take on active roles and help set direction. Women farmers in particular must be given a priority place. This is not for reasons of "political correctness" but because of women's often intimate knowledge of crop production and reproduction, their needs and interests in food security, and their leading roles in households, extended families, and social networks.

Breeders should do more work *in situ* — on farms and in communities — with farmers as colleagues, each complementing the other's knowledge, skills, and experience.

Decentralization should replace centralization as the main organizing principle to address specific local contexts.

Breeders should also collaborate with social scientists in an interdisciplinary research mode that takes into account both the biophysical and the social dimensions of the dynamic processes involved in maintaining diversity.

3. Pay attention to quality interaction and reward cooperation: farmers and scientists sharing benefits and costs

More evidence is required across regions and in specific contexts, but there is certainly enough to confirm that the participatory approach is both effective and efficient. **It requires a different**

way of organizing time, labour, and the research process with more emphasis on face-to-face interaction, especially in the field.

Start-up periods are usually very labour intensive, requiring a good deal of time and effort to lay a foundation of trust and to build working relationships. Longer term commitments are important to be able to create meaningful and effective collaboration and to cope with unavoidable setbacks, such as crop failure as a result of drought.

Facilitation and convening are new and important roles for researchers to take on. Additional training is an important investment if staff are lacking these skills. Working with a diverse group of people — including scientists in various fields, women and men farmers, and extension workers — means balancing a variety of ideas, interests, skills, and personalities. Managing the process of participatory planning, implementation, monitoring, and evaluation means paying significant attention to interactions and communications as well as ensuring openness and fairness. Building and strengthening the participatory process must be a central part of the agenda.

Participation in plant breeding also requires changes in how germplasm is selected, how experimental plots are designed, where experiments are implemented, and how assessment of the results takes place.

This method of organizing time and labour is called process management. **Promising and successful efforts must be recognized with incentives and rewards.** Farmers should be officially recognized as "coauthors" of new varieties or of publications that document the processes and final results. Breeders should be recognized and rewarded not only for the release of new varieties but also for their contribution to the process leading to the final products. Research grants should be targeted to proposals that deal adequately with process management questions.

4. Ensure good practice: five guiding principles

The principal aims of PPB are threefold — increased productivity, increased diversity, and the empowerment of farmers and other stakeholders. To contribute to these aims we need to track how and to what degree the methodology and particular tools influence both the effectiveness of research efforts and the empowerment of local resource managers. This requires a clear understanding of the types of learning that guide the process, and of the many variables that influence participatory practice. **Good practice means both contributing to local impacts and generating valid, trustworthy, and relevant research findings.** Relevance implies that these findings can be generalized; that they contribute to learning that can be applied in some way to areas beyond the research site.

The principles of good practice are as follows:

➤ The research reflects a clear and coherent common agenda or set of priorities among stakeholders, and it contributes to building partnerships.

➤ The research addresses the complex dynamics of change in human and natural resource systems and processes, and attempts to form an understanding of these, especially at the local level.

➤ The research links together various knowledge domains and applies the "triangulation principle" of encompassing multiple sources of information and methods.

➤ The research contributes to a concerted planning effort for social change.

➤ The research process is based on iterative learning and feedback loops, and there is a continuous two-way sharing of information.

5. Assess results through participatory monitoring and evaluation

Addressing users' needs and interests requires a different way of identifying good results. No longer can we rely just on predefined criteria generated in places far removed from farmers' fields and realities. No longer can we accept only the views and judgements of scientists or managers. New monitoring mechanisms, tools, and indicators are needed that reflect the dynamic, collaborative, multistakeholder nature of PPB.

Participatory monitoring and evaluation (PME) is an approach that opens a new window on research practice. It brings researchers together with other stakeholders — such as farmers, government officials, and extension workers — to monitor and evaluate research or development activities. Integrating PME into the project cycle strengthens the learning, accountability, and effectiveness of research efforts, in particular through the realization that **what matters is not only what is assessed but also who does the assessing.** PME contributes to a better understanding of how different concerns and interests are represented and negotiated.

PME also contributes greatly to a better understanding by researchers and local government staff of the interests and needs of both women and men farmers. Broadening the involvement of the various stakeholders in identifying and analyzing change creates a clearer picture of what is really happening on the ground and can include the perspectives of women, men, and various age, class, and ethnic groups. It encourages people to share successes and learn from each other.

Useful tools include ranking diagrams of various kinds; strengths, weaknesses, opportunities, and threats (SWOT) analysis; impact diagrams and matrices; focus group discussions; and self-evaluation forms. Criteria and examples of indicators used to measure results include the effects on

➤ improved farmer production (increased yield, improved cooking quality of grains),

➤ increased farmer-held diversity (increased number of varieties per crop, drought-tolerant varieties introduced),

➤ strengthened local organization of crop management and seed production (women take the lead in seed production and marketing, farmers organize local research groups),

➤ more dynamic and participatory formal breeding process (breeders have a better understanding of farmers' criteria),

➤ more dynamic and integrated organization of formal breeding and seed production (decentralization of experiments, landraces collected for inclusion in breeding programs), and

➤ empowerment (farmers ask breeders to extend PPB to other crops, farmers train other farmers).

6. Nourish a new generation of practitioners: innovate teaching and training methods

Learning by doing is useful to create a pool of experienced and knowledgeable practitioners in public and private organizations. However, new training and teaching methods must be designed, executed, and assessed to enlarge this pool and to speed the uptake of the approach. **Natural scientists (plant breeders and agronomists), social scientists (economists, sociologists, and anthropologists), and lawyers must broaden their knowledge and skills base across disciplines.** They need to learn to work together and be able to better complement each other. They need to be able to use appropriate participatory methods. They should be able to bring together various stakeholders and facilitate fluid and ongoing communication and cooperation.

Short, iterative training courses can be a means to acquire the new knowledge and skills, but more fundamental changes in graduate and postgraduate programs are necessary to train the future managers in research and policy.

The challenge

Affirmative action in these six areas is needed now to build on what has already been achieved, not only by IDRC but also by like-minded organizations around the world. However, it is an unfortunate fact that policy-making processes tend to be slow, complex, and political in nature, whether at the local, national, or international level. If we are to conserve the world's biodiversity, we must find ways of overcoming that structural inertia. In the words of the Crucible Group:

> Policymakers must find a way to stimulate innovation at the community, national, and international levels — in formal and informal, public and private sectors. The challenge ... is to find equitable mechanisms that allow these diverse forms of innovation to collaborate for the benefit of humanity.
> — People, Plants, and Patents, IDRC 1994, p. 43

Part 6

A Vision for the Future

In the first four parts of this book we examined some of the issues surrounding the loss of agrobiodiversity, reviewed 10 years of support for agrobiodiversity and PPB research around the world, and described some significant achievements as well as what has been learned as a result of this effort. It has also become clear that there is still much to be done and a great deal more to be learned.

In response to this need, the preceding chapter presented a series of six recommendations for action:

➤ Action on a global scale that would bring about the changes that are needed to support the processes that maintain diversity;

- Action that would break down the barriers between research "in the lab" and experimentation in the farmers' fields;

- Action that would build on this base of knowledge and experiences to create a greater awareness of the importance of conserving agrobiodiversity and bring about widespread implementation of the PPB methodology.

Recommendations, of course, are just words. What matters is implementation, and that represents a major political challenge. In this final chapter let's extrapolate from those recommendations. **What follows is a speculative look at how things could be 10 years from now.** In this journey into the future we will assume that NGOs, NARS, CGIAR, the policymakers, and the donors found the will and the resources to implement these recommendations. It is now 2012. Let's revisit the six recommendations presented in Part 5 to see where they might lead.

Increased relevance

> In China, the involvement of agricultural policymakers and policy institutions in the project is crucial. Through their involvement they can see for themselves that things can be done in a different way.
> — Yiching Song (project leader, China, 2002)

Seeds are no longer an afterthought
In the China of 2012, agricultural policymakers and policy institutions have been actively involved in efforts to conserve biodiversity. As a result, the critical importance of conserving agricultural biodiversity is widely accepted, and PPB has been embraced as a new and rational way of improving crops and

increasing plant genetic diversity. **Equally important, PPB has also been accepted as a new way of doing research.** But it is understood that PPB efforts on their own cannot be sustained if the context in which they operate does not provide the space and support. Therefore, society's ways of thinking about how food is produced and biodiversity is maintained has undergone dramatic change. "Taking care of the land" has been accepted as the guiding norm by all those involved in the food-production chain, from farmers and researchers to processors and policymakers. As a result, production has increased — particularly in those regions where it is needed most — and a continuous process is underway to innovate agricultural sciences, technologies, and policies.

PPB and biodiversity conservation have been bridged with sustainable agriculture and rural development. PPB has been taken out of its box. The focus is no longer solely on crops and crop diversity but also encompasses the people — the women and men whose knowledge, skills, and adaptive management practices maintain **and depend on** the variety of agricultural resources on-farm and off-farm.

Perhaps most significantly, **seeds are no longer seen as an afterthought**. PPB is one side of the coin, seed production and exchange systems are the other; they are inseparable. In this vision of the future it is broadly accepted that PPB, vital as it is, can be sustained only if there are viable and dynamic, local and national seed systems. In Guangxi, for example, numerous small companies have sprung up, many of them led by women. They stock and sell a much wider range of maize landraces and open-pollinated varieties than was available just a few years earlier.

New partnerships

The Farmers Research Committee is one of the unique groups involved in varietal development activities in Nepal. During visits to the research station before their involvement in varietal development, they criticized the researchers. But with 3 or 4 years of exposure to the research process, their understanding has been increased.
— Sanjaya Gyawali (researcher, Nepal, 2002)

Everything connects

In this new environment where PPB is accepted as the norm, it is only natural that local community-based agrobiodiversity conservation and improvement activities are connected to changes at the international and national policy levels. Thus there is opportunity for community input to global arrangements such as the CBD, the FAO's International Treaty on Plant Genetic Resources, and the WTO's agreement on the trade-related aspects of IPR. In this way the global context supports the diversity of local efforts, and the local diversity informs and guides the global.

Because it was a pioneer in this field, Nepal in 2012 is a leader and seen by many as an example to follow. Farmers' committees, made up of about equal numbers of women and men, now work closely with the formal sector in developing and evaluating new varieties, and in testing postharvest technology. Recognition by the government of farmers' rights has not only brought new respect to rural communities, it has also raised the level of participation in community affairs and improved local economies. Biodiversity fairs are popular and well attended, and the winners at these fairs are invited to become members of local and regional variety-release committees.

Quality interaction and cooperation

Even though I am a strong believer in the quality of science in PPB, *the key factor in the success of a project is the establishment of good human relationships, and this in turn is based on respect. I have learned that this is the major difficulty that some* NARS *and international scientists have.*

— Salvatore Ceccarelli (project leader, Syria, 2002)

Recognizing farmers' contributions

Given the broad-based acceptance of PPB, it is accepted as a normal state of affairs in 2012 that researchers, extension agents, and farmers — as well as other stakeholders, such as processors and traders — work together side by side. They all are making better use of researchers' access to knowledge, from breeding principles and methodologies to seeds and technologies to social science insights. All these stakeholders rely more on farmers' know-how, management, and organizational capacities to be able to address more precisely the needs and interests of a broad range of users, differentiated through factors such as gender, class, age, and ethnicity.

As a result of changing attitudes, ethical issues and intellectual property rights are a standard and important part of the research and policy agenda and discussed **from the very beginning** of new initiatives. Many issues that were previously ignored are now raised and dealt with. These include prior informed consent, explicit *ex ante* defined access- and benefit-sharing agreements, recognition of farmers' contributions to the creative process, and recognition of the farmer's rights to distribute, exchange, or sell seeds. These issues have been incorporated in policies and practices of CGIAR centres, NARS, and NGOs, and are part of the curriculum in training programs and teaching courses.

Good practice mainstreamed

The main achievements to date are the changes taking place in the plant breeding structure and process — to arrive at a more farmer-oriented science. These changes are for the longer term, not just for a project cycle. They include the decentralization of testing to off-farm sites, on-farm trial designs which are interpretable by farmers and which are conducted under "real farmer" input levels, and serious use and integration of farmer evaluations.
— Louise Sperling (former PRGA/PPB coordinator, 2002)

Building on success

Because of the increased acceptance of and interest in PPB, documentation and analyses of longer term diversity trends are more readily available. Toward the end of 2012, a workshop on PPB practices is attended by more than 150 researchers, research managers, farmers, and government officials from around the world. A few of the participants recall a similar, much smaller workshop in 2002 at which promising signs of progress in the PPB field were reported. At this gathering 10 years on it is reported that sound social analysis is now common practice in many countries. Researchers and policymakers pay systematic attention to resource tenure and its links with diversity and the livelihoods of rural people, especially the rural poor. Several participants report that they now routinely include analyses of power relationships, organizational processes, and policy-making processes as an integral part of biodiversity research projects.

An agriculture ministry representative speaks of the impact of PPB on policy-making, and emphasizes the need for government policies to remain flexible and informed by on-the-ground realities. The key, she says, is for adaptive, participatory research and natural resource management approaches to allow the custodians of biodiversity to deal more effectively with heterogeneous and changing agroecosystems.

A group of farmers and researchers from Cuba report that legislation designed specifically to recognize farmers' contributions to crop diversity and improvement, and that guarantees fair access to diversity and fair benefit sharing, has recently been put in place and is being respected. This legislation is already having a positive effect on the lives of farmers nationwide.

Quality participation

Women welcome the invitation to take part in a CIAL *in their community. I think this positive response has to do with their perception of a* CIAL *being a window to improving their livelihood conditions. They also think that being part of a* CIAL *allows them to show their skills, capacity, and potential; and in this way they are able to contribute to solving the problems of their community.*
— Noemi Espinoza (researcher, Nicaragua, 2002)

Gaining respect and influence

Across Latin America in the year 2012, CIALs are now more than a movement, in many countries they are part of, and supported by, the Ministry of Agriculture. CIAL representatives are respected and influential members of provincial and national policy advisory bodies in Nicaragua and a number of other nations. Governments as well as national and international institutions allocate resources to ensure that PPB is central to all agricultural research and policy efforts.

In the field, monitoring and evaluation are no longer the sole prerogative of the researchers. This important task is now carried out with the active participation of farmers and other stakeholders, who monitor and evaluate research or development activities. This has come about in large part because of the widespread adoption of the PME process, which has enhanced both the quality and the reach of people's participation.

The CIAL is generally a much more inclusive organization and welcomes those who for a long time were not meaningful participants — particularly women and members of poorer households. This appears to have come about as a result of careful reflection on decision making in the various steps of the research and development cycle. Collaborative forms of participation, decision-making, and agenda setting are now common practice.

Perhaps most significantly, the CIAL model has "traveled" to other part of the world. In Asia and Africa, and even in some countries of the North, communities are forming their own versions of the "local agricultural research committee" to gain a greater measure of control over their biodiversity and their livelihoods.

An active new generation of practitioners

What we want to get to is the training of professionals who are capable of working with both the natural and the social sciences.
— Humberto Ríos Labrada
(INCA project leader, Cuba, 2002)

Training tomorrow's leaders

Last, but definitely not least, the widespread acceptance of PPB as a new research methodology has made available the resources needed to develop new teaching and training methodology and materials that meet the demand for more and better training. An annual PPB course in Cuba attracts researchers from all over the world. In 2012, PPB has caught the imagination of a new generation of young professionals who want to get involved in the worldwide effort to conserve biodiversity. They will be the researchers and managers of tomorrow — key players and willing participants finally in quantities essential to sustaining agro-biodiversity research programs.

Glossary of Terms and Acronyms

Agricultural biodiversity: Also called agrobiodiversity. The variety and variability of animals, plants, and microorganisms used directly or indirectly for food and agriculture (crops, livestock, forestry, and fisheries). It comprises the diversity of genetic resources (varieties, breeds, etc.) and species used for food, fuel, fodder, fibre, and pharmaceuticals.

AREA: Agricultural Research and Extension Authority (Yemen)

BMZ/GTZ: Federal Ministry for Economic Cooperation and Development/German Agency for Technical Cooperation

CBD: Convention on Biological Diversity

CBDC: Community Biodiversity Development and Conservation program

CCAP: Center for Chinese Agricultural Policy

CGIAR: Consultative Group for International Agricultural Research

CIAL: local agricultural research committee

CIAT: International Center for Tropical Agriculture

CIMMYT: International Maize and Wheat Improvement Center

Cultivar: See *Plant variety.*

Decentralized plant breeding (program): A well-defined set of breeding experiments carried out in a variety of local sites (communities, farmer fields) that represent real farming conditions, as opposed to a single, central, research station site that does not represent a real farming context.

Ecosystem resilience: The capacity of an ecosystem to withstand dramatic impacts. An ecosystem is the dynamic complex of microorganisms, plants, and animals including human communities and their nonliving environment, interacting as a functional unit.

Ex situ conservation: Literally conservation "off-site." The conservation of a plant outside of its original or natural habitats, such as in a gene bank (a facility where temperature and humidity are artificially controlled) or botanical garden and stored as a seed, tissue, entire plant, or pollen.

Experimental lines: A group of individuals of a common ancestry and more narrowly defined than a strain or variety. A pure line is a clone. In plant breeding, "line" refers to any group of genetically uniform individuals formed from a common parent.

FAO: Food and Agricultural Organization of the United Nations

Farmers' rights: The recognition of farmers (past, present, and future) as *in situ* agricultural innovators who collectively conserve and develop agricultural genetic resources around the world. As such, farmers are recognized as innovators entitled to intellectual integrity and to compensation whenever their innovations are commercialized.

Gene bank conservation: See *Ex situ conservation*.

Genetic erosion: The loss of genetic diversity within a population of the same species, the reduction of the genetic base of a species, or the loss of an entire species over time.

GIC: farmer experimental group (Cuba)

GMRI: Guangxi Maize Research Institute (China)

GRPI: Genetic Resources Policy Initiative

Hybrid (general): The first-generation progeny of a cross between two different parents. An intermediate plant resulting from the crossing of two or more different bioytypes of the same species or biotypes from two different species.

IARC: international agricultural research centre

ICARDA: International Center for Agricultural Research in the Dry Areas

ICRISAT: International Crops Research Institute for the Semi-Arid Tropics

IDRC: International Development Research Centre

In situ conservation: Literally "on-site" conservation. The conservation of plants or animals in areas where they developed their distinctive properties: in the wild or in farmers' fields. Compare to *ex situ* conservation.

INCA: National Institute for Agricultural Sciences (Cuba)

Intellectual property rights: Laws that grant monopoly rights to those who create ideas or knowledge. There are five major forms: patents, plant breeders' rights, copyright, trademarks, and trade secrets.

IPGRI: International Plant Genetic Resources Institute

IPR: intellectual property rights

Landrace: A farmer-developed variety of a crop plant that is heterogeneous, adapted to local environmental conditions, and has its own local name.

LI-BIRD: Local Initiatives for Biodiversity, Research and Development (Nepal)

NARS: national agricultural research systems

NGO: nongovernmental organization

OAU: Organization of African Unity

On-farm conservation: See *In situ conservation.*

Open-pollinated variety: A variety multiplied through random fertilization; as opposed to a *hybrid* variety.

Participatory plant breeding: Broadly defined as approaches that involve close collaboration between researchers and farmers, and potentially other stakeholders, to bring about plant genetic improvements within a species. PPB covers the whole research and development cycle of activities associated with plant genetic improvement: identifying breeding objectives, generating genetic variability or diversity, selecting within variable populations to develop experimental materials, evaluating these materials (this is known as *participatory variety selection*, or PVS), release of materials, diffusion, seed production and distribution. It also could include assessing existing policy or legislative measures, or both, and designing new ones where needed. Farmers and breeders, and other stakeholders — such as traders, processors, and consumers — can take on different roles at various points in the cycle but they join forces to bring about change.

Participatory variety selection: The selection of fixed lines (including landraces) by farmers in their target environments using their own selection criteria. Consists of four methodological steps: (1) situation analysis and identification of farmers' varietal needs, (2) search for suitable genetic materials, (3) farmers' experimentation with new crop varieties in their own fields and with their own crop-management practices, and (4) wider dissemination of farmer-preferred crop varieties.

Plant (geno)type: The entire genetic constitution of a plant variety. An *off-type* is a plant differing from the variety in morphological or other traits.

Plant breeders' rights: See *Intellectual property rights*.

Plant species: A group of organisms capable of interbreeding freely with each other but not with members of other species. In taxonomic classification, a subdivision of a genus; a group of closely related individuals descended from the same stock.

Plant variety: In classical botany, a variety is a subdivision of a species. An agricultural variety is a group of similar plants that by structural features and performance can be identified from other varieties within the same species. Synonymous with *cultivar*.

PME: participatory monitoring and evaluation

PPB: participatory plant breeding

PRGA: Participatory Research and Gender Analysis program (of the CGIAR)

PROFRIJOL: Central American bean network

PVS: participatory variety selection

SDC: Swiss Agency for Development and Cooperation

SEARICE: Southeast Asia Regional Initiatives for Community Empowerment

SGRP: Systemwide Genetic Resources Programme (of the CGIAR)

SUB: Sustainable Use of Biodiversity program (of IDRC)

SWOT: strengths, weaknesses, opportunities, and threats (analysis)

TAC: Technical Advisory Committee (of the CGIAR)

UNCED: United Nations Conference on Environment and Development

Variety release: The official approval of a variety for multiplication and distribution.

WTO: World Trade Organization

Sources and Resources

The focus of this book is IDRC's support for research on agricultural biodiversity. For those interested in learning more about the topic in general, there is a great deal of literature, both printed and on the Internet. This appendix offers a selection of resources for further study. The list is structured in the same way as the book, providing resources for each chapter.

This book is also an integral part of IDRC's thematic Web dossier on participatory plant breeding: **http://www.idrc.ca/seeds**. The full text of the book is available online and leads the reader into a virtual web of resources that explores a decade of research on agrobiodiversity and PPB.

The Issue

For a useful general overview of agricultural biodiversity world-wide, country profiles, qualitative as well as quantitative descriptions of diversity trends (dynamics, causes, and consequences), trade, access, and benefit issues, the following publications are recommended:

Dutfield, G. 2002. Intellectual property rights, trade and biodiversity. Earthscan, London, UK.

FAO (Food and Agricultural Organization of the United Nations). 1998. The state of the world's plant genetic resources for food and agriculture. FAO, Rome, Italy (available in print and on CD-ROM); see also FAO's Web site on plant genetic resources: **http://www.fao.org/ag/cgrfa/pgr.htm**

Fowler, C.; Mooney, P. 1990. Shattering: food, politics, and the loss of genetic diversity. University of Arizona Press, Tucson, TX, USA; see also the Web site of the Action Group on Erosion, Technology and Concentration (ETC), formerly the Rural Advancement Foundation International (RAFI): **http://www.etcgroup.org**

Posey, D.A.; Dutfield, G. 1996. Beyond intellectual peoperty: toward traditional resource rights for indigenous peoples and local communities. IDRC, Ottawa, Ottawa, ON, Canada.

Pretty, J. 2002. Agri-culture: reconnecting people, land and nature. Earthscan, London, UK.

Robinson, R. 1995. Return to resistance: breeding crops to reduce pesticide dependence. IDRC, Ottawa, ON, Canada.

Secretariat of the Convention on Biological Diversity. 2001. Global biodiversity outlook. Secretariat of the Convention on Biological Diversity, Montreal, QC, Canada; see also the Secretariat's Web site: **http://www.biodiv.org**

Ten Kate, K; Laird, S.A. 1999. The commercial use of biodiversity. Earthscan, London, UK.

Thrupp, L.A. 1998. Cultivating diversity: agrobiodiversity and food security. World Resources Institute, Washington, DC, USA; more information can be found at the World Resources Institute's Web site: **http://www.wri.org**

The Approach

Details about IDRC's Sustainable Use of Biodiversity (SUB) program, and the research projects it supports, can be found by selecting the **Biodiversity** link at IDRC's networking Web site: **http://network.idrc.ca.** Amongst the many relevant resources accessible from this site is the following review of IDRC's experience in supporting agrobiodiversity research:

Vernooy, R. 2001. Harvesting together: the International Development Research Centre's support for research on agrobiodiversity (results and challenges). SUB Program, IDRC, Ottawa, ON, Canada.

Experiences from the Field

Extended case studies, research papers, IDRC *Reports* articles, and other related resources can be found by following the "Case Studies" string at **http://www.idrc.ca/seeds**.

Participatory barley improvement in North Africa and the Middle East

The projects' Web site is **http://www.icarda.cgiar.org/ Participatory/FarmerP.htm**. The site presents a number of research documents and an excellent overview of ICARDA's work in "farmer participation."

Ceccarelli, S.; Grando, S.; Booth, R.H. 1996. International breeding programmes and resource-poor farmers: crop improvements in difficult environments. *In* Eyzaguirre, P.; Iwanaga, M., ed., Participatory plant breeding. Proceedings of a workshop on participatory plant breeding, 26–29 July 1995, Wageningen, Netherlands. IPGRI, Rome, Italy. pp. 99–116.

Ceccarelli, S. 2000. Decentralized participatory plant breeding: adapting crops to environments and clients. *In* Proceedings of the 8th International Barley Genetics Symposium, 22–27 October 2000, Adelaide, Australia. Department of Plant Science, Adelaide University, Glen Osmond, Australia. Vol. I, pp. 159–166.

Ceccarelli, S.; Grando, S.; Tutwiler, R.; Baha, J.; Martini, A.M.; Salahieh; Goodchild, A.; Michael, M. 2000. A methodological study on participatory barley breeding. I. Selection phase. *Euphytica*, 111, 91–104.

Maize improvement in southwest China
CCAP's web site is **http://www.ccap.org.cn**

Song, Y. 1999. "New" seed in "old" China: impact of CIMMYT collaborative programme on maize breeding in southwestern China. Wageningen Agricultural University, Wageningen, Netherlands.

Song, Y. 2003. Linking the formal and informal systems for crop development and biodiversity enhancement. *In* Conservation and sustainable use of agricultural biodiversity: a sourcebook. CIP-UPWARD, Los Baños, the Philippines.

Reshaping agriculture in Cuba
Detailed information on this project can be found at **http://www.programa-fpma.org.ni**.

Ríos Labrada, H. 2003. Farmer participation and access to agricultural biodiversity: responses to plant breeding limitations in

Cuba. *In* Conservation and sustainable use of agricultural biodiversity: a sourcebook. UPWARD, Los Baños, the Philippines.

Local agricultural research committees in Latin America
Detailed information on the CIALs of Latin America can be found at CIAT's Web site: **http://www.ciat.cgiar.org/ipra/ing/**.

Ashby, J.A.; Braun, A.R.; Gracia, T.; Guerrero, M.P.; Hernández, L.A.; Quirós, C.A.; Roa, J.I. 2000. Investing in farmers as researchers: experience with Local Agricultural Research Committees in Latin America. CIAT, Cali, Colombia.

Humphries, S.; González, J.; Jiménez, J.; Sierra, F. 2000. Searching for sustainable land use practices in Honduras: lessons from a programme of participatory research with hillside farmers. Overseas Development Institute, London, UK. AgREN Network Paper No. 104. **http://www.odi.org.uk/agren/papers/ agrenpaper_104.pdf.**

Vernooy, R.; Baltodano, M.E.; Beltrán, J.; Espinoza, N.; Tijerino, D. 2001. Towards participatory management of natural resources: experiences from the Calico River watershed in Nicaragua. *In* Lilja, N.; Ashby, J.A; Sperling, L., ed., Assessing the impact of participatory research and gender analysis. Program for Participatory Research and Gender Analysis, CIAT, Cali, Colombia. pp. 247–262.

Enriching maize and rice in Nepal
LI-BIRD's Web site is **http://www.libird.org**.

PRGA (Program for Participatory Research and Gender Analysis). 2001. An exchange of experiences from South and Southeast Asia. Proceedings of the international symposium on participatory plant breeding and participatory plant genetic resource enhancement. PRGA, CIAT, Cali, Colombia.

From this book, the following three papers emanate from the
Nepalese research project:

Subedi, A.; Joshi, K.D.; Rana, R.B.; Subedi, M. 2001. Participatory
plant breeding in diverse production environments and institutional settings: experience from a Nepalese NGO. pp. 75–86.

Joshi, K.D.; Sthapit, B.R.; Witcombe, J.R. 2001. The impact of
participatory plant breeding (PPB) on landrace diversity: a case
study for high-altitude rice in Nepal. pp. 303–310.

Subedi, M.; Shrestha, P.K.; Sunwar, S.; Subedi, A. 2001. Role of
farmers in setting breeding goals. pp. 311–318.

The global program on Participatory Research and Gender Analysis

The PRGA program maintains a very resource-rich Web site:
http://www.prgaprogram.org. Documents available on this
site include the recently pulished PRGA 5-year Synthesis Report and
the program's Annual Report, 2002. (**http://www.prgaprogram.org/
progress.htm**).

Eyzaguirre, P.; Iwanaga, M., ed. 1996. Participatory plant breeding. Proceedings of a workshop on participatory plant breeding,
26–29 July 1995, Wageningen, Netherlands. IPGRI, Rome, Italy.

PRGA (Program on Participatory Research and Gender Analysis).
1999. Crossing perspectives: farmers and scientists in participatory plant breeding. PRGA, CIAT, Cali, Colombia.

Learning from Experience, Recommendations, and Future Vision

The following publications present edited collections of crop-
conservation and crop-improvement studies from around the
world. The studies were funded by a variety of donors, including
IDRC.

Almekinders, C.; de Boef, W., ed. 2000. Encouraging diversity: crop development and conservation in plant genetic resources. ITDG Publishing, London, UK.

Brush, S.B., ed. 2000. Genes in the field: on-farm conservation of crop diversity. Lewis Publishers, Boca Raton, FL, USA; IPGRI, Rome, Italy; IDRC, Ottawa, ON, Canada.

CIP-UPWARD (International Potato Center, User's Perspectives with Agricultural Research and Development). 2003. Conservation and sustainable use of agricultural biodiversity: a sourcebook. CIP-UPWARD, Los Baños, Philippines.

Cleveland, D.A.; Soleri, D., ed. 2002. Farmers, scientists and plant breeding: integrating knowledge and practice. CABI Publishers,Wallingford, UK.

Cooper, H.D.; Spillane, C.; Hodgkin, T. 2001. Broadening the genetic bases of crop production. CABI Publishers, Wallingford, UK; FAO, Rome, Italy; IPGRI, Rome, Italy.

Crucible Group. 1994. People, plants, and patents: the impact of intellectual property on trade, plant biodiversity, and rural society. IDRC, Ottawa, ON, Canada. Available online: **http://www.idrc.ca/booktique**.

Crucible II Group. 2000. Seeding solutions. Volume 1: Policy options for genetic resources (*People, plants, and patents revisited*). IDRC, Ottawa, ON, Canada; IPGRI, Rome, Italy; DHF, Uppsala, Sweden. Available online: **http://www.idrc.ca/booktique**.

———— 2001. Seeding solutions. Volume 2. Options for national laws governing access to and control over genetic resources. IDRC, Ottawa, ON, Canada; IPGRI, Rome, Italy; DHF, Uppsala, Sweden. Available online: **http://www.idrc.ca/booktique**.

de Boef, W.; Amanor, K.; Wellard, K.; Bebbington,. A., ed. 1993. Cultivating knowledge: genetic diversity, farmer experimentation, and crop research. ITDG Publishing, London, UK.

Friis-Hansen, E.; Sthapit, B., ed. 2000. Participatory approaches to the conservation and use of plant genetic resources. IPGRI, Rome, Italy.

Jarvis, D.; Sthapit, B.; Sears, L., ed. 2000. Conserving agricultural biodiversity *in situ*: a scientific basis for sustainable agriculture. IPGRI, Rome, Italy.

Partap, T.; Sthapit. B., ed. 1998. Managing agrobiodiversity: farmers' changing perspectives and institutional responses in the Hindu Kush-Himalayan region. International Center for Integrated Mountain Development, Kathmandu, Nepal.

Prain, G.; Bagalanon, C.P., ed. 1998. Conservation and change: farmer management of agricultural biodiversity in the context of development. UPWARD, Los Baños, Philippines.

Sperling, L.; Loevinsohn, M., ed. 1997. Using diversity: enhancing and maintaining genetic resources on-farm. IDRC, Ottawa, ON, Canada. Online only: **http://www.idrc.ca/library/document/ 104582/**.

More information

More information about some of the projects and organizations supported by IDRC and mentioned in this book can be found at the following Web sites:

http://www.cdbcprogram.org: The Community Biodiversity Development and Conservation program is a global initiative aimed at supporting the ongoing efforts by farming communities to conserve and develop agricultural biodiversity.

http://www.cimmyt.cgiar.org: The International Maize and Wheat Improvement Center conducts research on maize and wheat to help people overcome hunger and poverty and to grow crops without harming the environment. CIMMYT is one of the 16 IARCs.

http://www.icrisat.cgiar.org: The mandate of the International Crops Research Institute for the Semi-Arid Tropics, one of the 16 IARCs, is to enhance the livelihoods of the poor in semi-arid farming systems through integrated genetic and natural resource management strategies.

http://www.ipgri.cgiar.org: The mandate of the International Plant Genetic Resources Institute is to advance the conservation and sustainable use of plant genetic resources. IPGRI is one of the 16 research institutions that make up the CGIAR.

http://www.searice.org.ph/programs.htm: The Southeast Asia Regional Initiatives for Community Empowerment is a regional research and advocacy NGO focusing on community-based conservation, development, and utilization of plant genetic resources.

The Publisher

The International Development Research Centre is a public corporation created by the Parliament of Canada in 1970 to help researchers and communities in the developing world find solutions to their social, economic, and environmental problems. Support is directed toward developing an indigenous research capacity to sustain policies and technologies developing countries need to build healthier, more equitable, and more prosperous societies.

IDRC Books publishes research results and scholarly studies on global and regional issues related to sustainable and equitable development. As a specialist in development literature, IDRC Books contributes to the body of knowledge on these issues to further the cause of global understanding and equity. The full catalogue is available at **http://www.idrc.ca/booktique**.